WHISKIES
GALORE

Whiskies Galore

*A Personal Journey
(of Sorts, and a Sort of a Journey)
around Scotland's Island Distilleries*

Ian Buxton

BIRLINN

First published in Great Britain in 2017 by

Birlinn Ltd
West Newington House
10 Newington Road
Edinburgh
EH9 1QS

www.birlinn.co.uk

ISBN 9781780274423

British Library Cataloguing-in-Publication Data
A catalogue record for this book is available
on request from the British Library

Text design by Jules Akel

Printed and bound by CPI Mackays, Chatham, Kent

Dedication

THIS BOOK is dedicated to my mother and father, who first took me to Scotland's islands. Not that I had any choice in the matter, and they were conscientious enough not to leave me there—tempted though they no doubt were. But thanks anyway. And to my wife, who has put up with my usual distracted nature and inability to engage in matters domestic while I was writing it. I'm told that wasn't easy.

CONTENTS

A different sort of journey

THIS IS AN ACCOUNT OF A PERSONAL JOURNEY. Now I don't know if you want to know where I was born, and what my perfectly agreeable middle-class childhood was like, and how my parents were occupied and all before they had me, and all that Holden Caulfield kind of angst, but we're here to talk about malt not rye, so I won't be going into it, if you want to know the truth.

To start with, although we shall encounter whiskies galore (and a few gins for good measure), this is not your run-of-the-mill 'whisky book'. That is to say it does not catalogue histories, at least not in any consistent way; capacities and outputs are recorded only intermittently; expression after expression is not laboriously listed and detailed tasting notes made, and there are very few artful photographs of grizzled chaps rolling barrels or serious-looking coves staring moodily at a tasting glass full of liquid illuminated by a fortuitous shaft of iridescent shimmering golden light. Grim industrial buildings have not been tastefully styled; neither have I romanticised dereliction, nor has an art director's soft-focus lens imposed spurious romance on a grimy and cobwebbed cellar. This is not whisky porn.

Why is that? Well, if what you want is detailed information on equipment, still sizes, barley varieties, output, capacities and so on and so forth, or exhaustively documented lists of 1,001 different whiskies with lengthy and baroque tasting notes, there are many places to find those doubtless fine and important things—whisky blogs, websites and books are there in splendiferous plenty, so enjoy. There seemed very little point in repeating what had already been so thoroughly well done by others, doubtless more panoptic in scope, conscientious in delivery and meticulous in presentation than me. At any event, a tasting note is only one person's opinion and an unreliable guide to the probability of your pleasure. Above all, avoid the absurdly spurious accuracy of a tasting score of 94.5 points or similar nonsense. Trust your own judgement; enjoy what you enjoy and don't let anyone else tell you otherwise.

Other people have compiled extensive lists and faithfully recorded output and so on, but when I started a similar inventory I very soon thought it pointless and unkind to impose another catalogue of such drudgery, however conscientious, on an un-suspecting world. Finally, and most importantly, that kind of book is redundant in the age of the internet. Chances are that by the time it is written, fact-checked, designed, proofread, printed and delivered to the shops it will be out of date and it will continue to get ever more out of date the longer it stays in print. The world of whisky changes very fast and this type of information is simply to be found faster and more reliably on the web.

However, I do find it interesting that in recent years Scotland's island distilleries have attained a special status in the eyes of many drinkers, and a romantic image has formed of island life and the drams produced there. Island whiskies, and to a lesser extent gin, continue to exercise a fascination and powerful emotional draw on consumers the world over. Their remarkable combination of heritage, mystique, remote location and, of course, the highly distinctive taste of many of them is often imitated but seldom bettered: they have a unique romance and attract visitors from round the world, especially Germany and Scandinavia. What is more, I have, quite by accident, joined a wedding party who had travelled from Singapore to be married and celebrate their union on Islay, so the appeal is clearly global.

So I set off with the loosely-defined hope that I might discover what Scotland's island distilleries are really about. Are they, as their enthusiastic disciples proclaim, really so different? Is there some elusive quality involved or some spirit of place or something so remarkable about the people who distil on an island that makes their products intrinsically different and superior? Or is that a self-serving myth, merely the creation of shrewd marketing experts looking to exploit any difference in a competitive world?

In the process, what emerged was an entirely impressionistic ramble through my memories of the islands going back some fifty years and my more recent experience of Scotland's island distilleries. Some are more important than others and may receive more space, but that will depend on my personal relationship with them and their island. It is, if you will, a highly personal, whisky-fuelled journey through those islands with frequent diversions on topics which happen to amuse, interest or divert me.

Some people collect whiskies but never drink them. Some people — may they get help soon — even promote whisky as an 'investment', a trend upon which certain distillers have happily capitalised, leading to ever more elaborately packaged special bottlings of increasing cost and vulgarity and a general drift upwards in the price of whisky. All this is to be deplored by the genuine and righteous student of whisky who understands instinctively that whisky was made to be drunk and has no meaning until the moment of its consumption.

I despair of the whisky investment hucksters and I fear the craze will end badly. However, I understand the urge of the collector: I have lots of pens; more pens than strictly speaking I 'need', but the difference is that I wrote this book with them. You will have to indulge me, therefore, as the pens and the ink with which I wrote this book make cameo appearances throughout the narrative that follows. This book features aficionados, collectors and obsessives; and I, in my own small way, am one of them.

While in confessional mode, I really should acknowledge that although I have spent nearly half a century visiting Scotland's patchwork of islands, it would be wrong of me, as an outsider, to cast grand judgements upon them. Instead I offer you my thoughts, random and incoherent as they most definitely are, on Scotland's island distilleries—to places, people and landscapes that have in common an existence, sometimes fragile, 'on the edge'—an edge physical, geographic, cultural and economic. And so, without further delay…

2

Arran

Look out! Shark!

THE GREAT, GREEN-GREY MONSTER SURFACED LAZILY right by our little boat, suddenly so flimsy and unstable and the coast line so far distant. Another few feet and we would have been over-turned and in all probability drowned, the behemoth by our side drifting lazily on, entirely oblivious to our cries and panic. Not waving but drowning.

But first, a few facts.[1]

Arran is the most accessible of islands. It lies in the Firth of Clyde, conveniently close to Glasgow, and is served by a frequent

---

1 You probably realise what I've just done: I open with a cunning hook to draw you in, with the promise of a subsequent juicy conclusion. What happens to Ian? Anxious to know, you read on. Actually, it's an ancient rhetorical device, much favoured by Cicero, known as *parekbasis*, or a digression from the main topic. Boris Johnson, a great classicist (and at the time of writing our Foreign Secretary), uses it all the time. You may now proceed: there will be a pay-off for the story later, though, as you will have deduced, I didn't actually drown.

car ferry which crosses from Ardrossan to the island's second largest town, Brodick. During the summer months, a smaller vessel travels from Lochranza in the north of Arran to Claonaig on the Kintyre peninsula. Apart from a slipway for the ferry traffic there isn't a lot to see there unless you're interested in bus shelters. I'm not, unless waiting for a bus in the rain, in which case I'm an avid enthusiast. Actually, there's also a very dramatic modern house there, overlooking the sea, owned by a thoroughly charming Swiss pension fund manager who, from time to time, invites me for tea. It's a long way to go; though, to be fair, even further from Zurich, so I have yet to enjoy his hospitality.

At 167 square miles, Arran is the largest island in the Firth of Clyde. Today, with a population of some 4,600, tourism and services to support tourism appear to be the main source of employment but, in its heyday (1821), Arran had around 6,600 inhabitants largely engaged in subsistence agriculture. However, extensive 'improvements' (aka Clearances) by the 10th Duke of Hamilton led to a significant drop in the number of inhabitants by the end of the nineteenth century. The unfortunate islanders mostly emigrated to Canada, where many had been promised land to farm. Arran's history of clearance is relatively little known compared to the more notorious Sutherland improvements, but doubtless nonetheless distressing for the victims.

Not everyone thought the Clearances a bad idea. In July 1851, The Scotsman wrote that 'Collective emigration is, therefore the removal of a diseased and damaged part of our population. It is a relief to the rest of the population to be rid of this part.'

No mention, you might note, of what the Canadians thought of their new would-be citizens. The migrants carried their Gaelic language and culture with them and one consequence of their removal was the effective extirpation of a Gaelic identity on Arran. James Hogg, amongst others, was heard to lament its loss. By the

2011 Census a mere 2% of Arran residents aged three and over could speak Gaelic, the last native speakers of Arran Gaelic having died in the 1990s.

One main road runs around the island and two across it, though one gains the impression that most visitors venture little further than either Lamlash or Brodick, both of which offer a slightly dilapidated tourism product, which is not without its fin-de-siècle charms. Decent coffee though, which is something relatively new and very welcome. There was little of the kind here in the mid-1980s when I was a frequent visitor to Arran Provisions' jam and mustard factory, part of an ill-fated diversification venture by my then employers, the blenders Robertson & Baxter, who seemed convinced that the glory days of whisky had passed and it was time to invest their profits in the food industry. Unfortunately, it seemed to have escaped their notice that Arran Provisions, though they manufactured food products, were in fact engaged in the gift business. In career terms, this led to considerable frustrations and not a few tears, though I got a very nice pen out of it and seldom went short of tasty preserves. Christmas gift-giving was, for a brief period, the least of my problems.

Arran enjoys a favourable climate, being sheltered from the worst western storms by Kintyre and benefiting from the Gulf Stream. However, it does rain a lot — at least it has done on the several occasions that I have been there. I feel it's probably personal.

The first occasion was in the mid-1960s when my family took a holiday on the island's west coast in a dank and dreary cottage near Pirnmill. One incident stands out in my memory. We had hired a small rowing boat to go out fishing for mackerel, only to be surprised and not a little perturbed by a large basking shark surfacing right beside the boat.

In my memory it was considerably longer than the boat, which is in fact entirely probable as a mature basking shark can measure more than thirty-six feet in length (for anyone under forty, that's eleven metres). It was a small boat, containing me, my father and a younger brother. We were not, of course, wearing life-jackets.

As Norman MacCaig has it, it was a monster that rose 'with a slounge out of the sea'. To a teenage boy this seemed immensely exciting; looked at in hindsight I tend to the poet's view that once, at such close proximity, was once too often. I now maintain that they are best observed from the shore or else, as another shark watcher would have it, you're going to need a bigger boat.

That allusion to the wise-cracking Brady of *Jaws* leads me to recall that then, and for some decades afterwards, there was a fishery taking basking sharks off Arran and other Hebridean islands. The naturalist Gavin Maxwell established a fishery on Soay, off Skye, shortly after the Second World War and the leading fisherman in the Clyde, Howard McCrindle, operated as late as 1997, finding the sharks by means of echo-location. He could catch over a hundred fish a year, with the deep water off Lochranza being a prolific hunting ground.

The basking shark fishery seems to have had periods of boom and bust, presumably due to over-exploitation of a vulnerable population. There is a fanciful engraving showing the taking of a basking shark off Lochranza in Thomas Pennant's *A Tour in Scotland* (1772) in which he gives a lively account of their capture. Herring proved more abundant, however, and easier and less hazardous to catch. An 1808 account, and one from 1820, both mention the herring fishery and by 1875 it was certain that the pursuit of the basking shark had been abandoned, presumably as uneconomic. Fishing resumed following the Second World War, but McCrindle's operation, be it ever so technologically sophisticated, was something of a throwback and heavily criticised in its later years by environmental activists.

Rightly so, it turns out, as live sharks are a great deal more valuable than dead ones. Today there is a burgeoning tourism business offering trips to see the basking shark at sea. Once, I felt, was often enough—no shark could compare with that of my childhood adventure and so, though tempted, I decided against even a short exploratory voyage, delighted though I was to learn that these fine fish are exploited today by the camera rather than harpoon.

Seen through a contemporary lens, Arran seems a trifle improbable as a location for a distillery. Though its publicity makes much play of a flourishing illicit distilling tradition, the evidence for that is largely anecdotal and thus both sparse and unreliable. Such a claim could, and indeed is, made for much of Scotland, Ireland and even Northumberland, not to mention the seedier parts of Boston in Lincolnshire where five unfortunate Eastern European workers died in 2011 following an explosion in their bootleg vodka distillery.

That would have been my normal, somewhat cursory response to the claim of a heritage built on illicit distilling—which, if we're honest, is not a particularly unusual or distinctive back story. However, Neil Wilson, an erstwhile colleague on various projects, has recently written a detailed account of the Arran distillery and in it he quotes various authorities and more recent academic research. From all that work we may conclude that there was indeed a flourishing trade in contraband spirit on Arran and, latterly, three licensed distilleries. These, however, failed in the first half of the nineteenth century, as did so many others across Scotland—possibly a warning to the latest generation of entrepreneurs convinced that whisky is their route to fame and fortune.

He further relates that bootleg distilling may well have continued late into the twentieth century, encouraged, perhaps, by the strictures and shortages of the Second World War. Could it continue to the present day in some remote glen? One hesitates to speculate.

What is certain is that there is occasional reference in Victorian literature to the quality of the 'Arran Water'. Aeneas MacDonald in *Whisky* quotes Strang's *Glasgow and its Clubs* (1856; it's ever at my shoulder) in which it is suggested that in the eighteenth century at least, Arran whisky was regarded as 'the real stuff' and was as highly regarded as that from Glenlivet. It was apparently 'dispensed with a sparing hand'. Glasgow clubs must have changed in the intervening years—this sounds more like Edinburgh hospitality.

The island's best documented legal distillery (and this is not a particular distinction as its history is partial at best) seems to have been at Lagg in the south of the island. There is a slight connection to present day operations, however, in that the current distillery plans to build a new 800,000 litre capacity plant there. This will produce a heavily peated single malt to compete with Islay and other west coast malts, while the existing Lochranza operations will continue with the present style of production.

Excited, as ever, by the call of the new I drove to Lagg to see the site of the new distillery. Or, to be precise, I was driven, as my long-suffering wife had elected to accompany me and was deputed as chauffeur. I believe she may think I enjoy myself on these trips.

Calling the previous day at the distillery, of which more later, we received directions to the site. 'Look for a blue pipe in a field,' we were told. 'You can't miss it, it's just past the turn for the nudist beach.' Of course: the ominous optimism of 'you can't miss it'.

We then drove around for some time, in heavy mist, looking for a blue pipe in a field. We passed many fields, though only one sign for a nudist beach, observing on the way a sadly abandoned and slightly forlorn church, its windows shuttered and doors bolted, where I immediately conceived a grand design to buy and restore the building as a showpiece home. We saw, briefly, some passing wildlife and a number of crows feasting greedily on a dead sheep. For some reason best known to herself and never revealed, Mrs

Buxton began reciting *The Ballad of Sir Patrick Spens*. Not only the climate but the soundtrack to our exploration was proving sombre.

We saw what appeared to be many molehills. I was amused to read recently that there are three professional bodies for mole catchers in the UK and that they are engaged in what I fear I must term a turf war. The British Mole Catchers Register is at odds with the Association of Professional Mole Catchers and both are criticised in turn by the Guild of British Mole Catchers. You'd think they would stop digging.

In a previous life I was mole catcher. That isn't actually strictly true, though I did once employ a mole catcher, an elderly gentleman who was in the habit of leaving the corpses on our doorstep and then arriving some days later to collect his fee. There is very little eating on a mole so I buried them, which seemed at once ironic and yet utterly appropriate given that if they had died naturally the deceased moles would have remained underground for all time.

Perhaps fortunately, for they usually disappoint, we did not see any nudists, though the sign to their beach was quite plainly evident. That was actually something of a surprise, as having visited the original distillery in bright sunlight, we were attempting to find its new stablemate's home in low cloud and a very heavy downpour. Indeed, visibility was so poor that it was hard to determine anything beyond the few hedgerows, which it would have been possible to see over had it not been for the thick mist and torrents of rain. What memories the rain brought back of long-forgotten trips to the west of Scotland.

Lacking serious storm gear, it was impossible to step out of the car to investigate possible sightings of blue pipes, and that had very little appeal, even if the necessary heavy-duty clothing had been available. It was, we assured ourselves, only a blue pipe in a field, and you can see one of those almost any day you choose. After a while we gave up and returned to look at the church, which

I resolved not to buy. We did, however, conclude that the new Arran distillery will not want for water.

It did strike me as more than a little perverse to go to such lengths and build at this distance. Lagg is about as far away from the original site at Lochranza as it is possible to be on a twenty-five-mile-long island—that is to say, around twenty-five miles—and the current distillery is hardly hemmed in by other developments.

In fact, it sits on open ground, surrounded by rolling acres of fairly unprepossessing fields which appear to the layman to be of little use to man nor beast. But the landowner, learning of the distillery's wish to expand, apparently disagreed and placed so forbidding a premium on the ground that any possibility of development was forever forestalled. So, to Lagg they must hie, and there the future lies.

For the present then, the Arran distillery is to be found at Lochranza, previously known best for its castle and summer car ferry to Claonaig. It is but twenty-one years old, making it one of the newest of the island distilleries and, in an industry where much credibility is attached to sheer age, and great ancestry is venerated with almost Asiatic reverence, that can hardly account for its appeal. But nearly 90,000 visitors passed through its doors last year, numbers that most distilleries in Scotland would be indecently happy to welcome.

Let us in a brief digression contemplate the matter of distilleries and their visitors. Not so very long ago, the idea of visiting a distillery for pleasure seemed a very strange one, particularly to an industry that was then primarily under the control of the blending and production function, with a lot of accountants in the background (I suspect they actually ran things, but let the distillers imagine they were in charge—life would have been quieter that way, and the traditional whisky industry was partial to a quiet life, interspersed with a decent lunch). With the exception of the Tobermory distillery, which we'll visit eventually, who were apparently encouraging visitors as far back as 1923,[2] generally speaking distilleries closed their doors to all but the most persistent or well connected.

It's fascinating to see what our forebears thought most important: I have in my library a charming little book, the *History of the House of Buchanan*, published in 1931. It is lavishly illustrated for the period and there are pictures of the head office; the Sydney office (and the bottle washing plant and labelling room in Sydney); the case works; the bottle works, filling machines and bottling store (all staffed by serious-looking females in drab overalls); the cooperage; the Glasgow warehouse; and even the Holborn Convivial Society Rifle Club Prize Distribution dinner (uniformly attended by serious-looking gentlemen in evening dress). In total, there are twenty-nine illustrations in a thirty-nine-page book. Of distilleries, there are but two—Anderson's Black Swan Distillery in Holborn (of Gordon Riots fame), purchased in 1898 and promptly remodelled as office and stores, and a distant view of the exterior of the Bankier Distillery (1903). This in a book recounting the history of a whisky company.

The key point is that distillery interiors were evidently considered of little or no importance. The sole insight into production is a view of some blending vats and casks in a mysteriously unidentified Blending Department (location not given). Considerably more space is given over to the company's horses and horse-drawn vans. Why, the book seems to ask, would anyone be interested in what went on in the still room? It's not as if it was considered a sacred and arcane priesthood with secret knowledge that could not be shared—the assumption was that no one was very interested.

---

2 Or possibly 1928; the date on the relevant photograph is unclear.

Such a view was widespread until comparatively recently. Working in the industry in marketing in the late 1980s I was obliged to apply well in advance to my colleagues in production if I intended to visit one of the company's distilleries, and such proposed visits were greeted with some suspicion. The general feeling was that distilleries belonged to the production department and no one, let alone the public, had any legitimate interest or business in stepping inside them. Simply dropping in was regarded as very poor form and my crusty colleagues were not slow to point out my breach of etiquette. They were also not slow to proffer their opinion, largely unfavourable, on the efforts of the marketing department.

This was all changing then, however, largely due to the influence of William Grant and Sons who had opened modest visitor facilities at Glenfiddich as early as 1969. This was hardly universally welcomed and, indeed, greeted with some scorn by many in the industry, even when sales through their shop became visible. Led by Glenfiddich though, eventually more distilleries did open their doors, albeit reluctantly at first.

The general policy was to encourage free admission, something which has long been abandoned almost everywhere, as the price of admission has soared to a staggering £1,000 per visitor for the ultimate in distillery experiences. Access to the distillery, or as it is now termed, the 'brand home', has now become a function of marketing and a central adjunct of brand promotion. What dyed-in-the-wool distillery types make of this is not recorded.

However, enough grumbling. Let's join the 90,000 visitors to Arran who, we may safely assume, are unique to that distillery as we have established it's the only one on the island. First, we have to get there.

We had stayed the previous evening in dingy rooms in Ardrossan, of which the less said the better. The accommodation was promoted by means of a large banner on the exterior wall, offering rooms 'from £35'. £35 would have been about right but our visit coincided with an important golf competition at nearby Royal Troon so the price had risen to £95, and nothing else was to be had.

Well, it was the Open apparently, though that meant little to me, apart from a more acutely personal appreciation of the economic impact of golf tourism. Accordingly, we boarded Messrs Caledonian MacBrayne's ferry (a concern known to one and all as CalMac) in a less than sunny mood. But it only takes a few minutes of a level sea, a bright morning and CalMac's full Scottish breakfast for the mood to brighten, and we disembarked in better humour.

The drive from Brodick to Lochranza is a pleasant and undemanding one, provided you are not in any hurry. It winds along the coast through various small villages until eventually, having crossed a small bridge over the River Easan Biorach in the glen of the same name (don't even try to pronounce this; it translates as 'the Glen of the Sharp Waterfalls'), you will see the distillery on your left-hand side.

As befits a relatively modern and purpose-built construction, everything is very neat and trim. Today it claims to be the number one attraction on the island though inevitably some guests arrive more to seek shelter from the rain than out of any profound interest in whisky. But admission charges are actually fairly modest, from as little as £3.50, and there is an excellent café with a wide range of snacks and meals on offer. As rain shelters go there are many worse.

Distilling on Arran had something of an uncertain start. The proposal for a distillery was first mooted in March 1991 by Harold Currie, a former senior executive with Pernod Ricard, assisted by David Hutchison, a Glasgow architect. As the early years of the distillery have recently been fully documented in Neil Wilson's commemorative work *The Arran Malt* I don't propose to repeat the details here.

Curiously, however, it is incomplete, for he does not relate my small contribution — which was to pour scorn on the proposal in a report I prepared for Highlands & Islands Enterprise (HIE). He merely notes laconically that in meeting with HIE 'the company failed to raise any interest in the project'. I'm not surprised, as perhaps he was unaware of the background. I expect so. Perhaps now the story can be told ... here goes, in any event.

Sometime in 1992, the project's backers presented HIE with 'an eighty-page, A3, landscape-format development proposal' for a £1.422 million project. This quickly grew to around £3 million, at which point my opinion was sought. Having left my previous employer, I was then working as an independent consultant and I was briefed by HIE on the basis of the development documentation and asked to provide a view on the strengths and weaknesses of the proposal.

I see from my files that I prepared a twenty-page response, with a number of appendices. If I have to express an opinion it still reads well today, but it was not positive. I identified what I saw as fundamental flaws and weaknesses in the plan, though suggested that certain aspects, in particular the projection for visitor income, were actually conservative.

It seemed to me then that though the proposal was 'an interesting and exciting one that has great emotional appeal and considerable economic benefit to Arran' there were a number of overriding strategic concerns. In particular, I suggested that there was an 'urgent requirement' for 'considerable further detail in the sales, marketing, distribution and general management aspects' of the proposal. I also considered it underfunded and the sales projections to be 'extremely optimistic'. Phew! It was a pretty damming verdict.

HIE did not support the investment at that stage, though later provided some funds for the visitor centre and warehouse development. The publicly stated reason was that the employee numbers projected did not justify the public-sector funds requested. Behind the scene, however, HIE were uncomfortably aware of the over-capacity then existing in the Scotch whisky industry, a fact which my report had highlighted. So, full disclosure: had it been up to me the Arran distillery would never have been built. I could not but be acutely aware of this as I strolled nonchalantly to the building, as I guessed my delightful, helpful hosts would know nothing of my perfidy or would at least be too polite to mention it.

They were distillery manager James McTaggart, an industry veteran, and Faye Waterlow, manager of the busy visitor centre. They could not have been friendlier or more obliging, opening every door and fully answering all my questions. I guessed they knew nothing of my earlier contribution to the project.

Arran, I thought, was a very perky distillery. By rights, it should probably never have been built and it has certainly passed through more than its fair share of financial vicissitudes during the past twenty-one years, which have been nothing short of eventful. Like so many small distillery projects it has required regular re-financing and the Currie family interest (the founder Hal Currie died in 2016) has been greatly reduced over the years.

As the exhaustive commemorative history makes clear, the company regularly made heavy losses and it was not until 2006 that a trading profit was achieved. In the following years the deficits returned and only by 2010 was consistent profitability achieved. Without the patient, long-term support of shareholder Leslie Auchincloss, now the principal owner of the business following his repeated injection of cash in the early years, the company would have long since failed (as, I believe, someone predicted).

The early releases of whisky were not, to be kind, especially memorable, largely due to the shortcomings of poor quality wood used for maturation. But things have improved greatly since then—the wood policy has been revised; the techniques of finishing advanced and, under McTaggart's steady hand, the quality of the distillate immeasurably improved.

James McTaggart joined Arran in September 2007 from Bowmore distillery on Islay where he had worked for some thirty years. But as we walked through the distillery, warehouse and briefly to the water source by that unpronounceable river, he surprised me with the news that he was a commuter. Does that strike you as improbable? It seemed so to me, but he assured me he made the weekly commute by car and ferry from Islay. That involves a short drive to the Lochranza ferry; thirty minutes or so on the ferry itself; then the longer trip on the boat to Islay (at least another three hours, allowing for check-in, embarkation and so on) and finally another drive across Islay to home, always assuming no weather delays.

His wife, he told me, felt she would never settle on Arran as it was so very different from Islay. And so it is: Arran is lush, where Islay is more open and sparse; Arran has long depended on tourism, something which affects the soul of a place, not always to the good, whereas Islay has only recently become popular as a visitor destination and the numbers remain well below those of its near neighbour. More fundamentally, Islay is quite clearly a Hebridean island where Arran is, on a busy summer day, almost a suburb of Glasgow. It has long been popular as a resort location for Glaswegians, some making short holiday visits, more fortunate others maintaining a second property on the island. Their characters are different and I readily understood why a deeply rooted Ileach would find it hard to settle on Arran.

Nonetheless, I was intrigued by the weekly travelling and later we gave James a lift to the Lochranza ferry and saw him to his car parked at West Loch Tarbert to catch the Islay boat, where he was immediately amongst friends. I do sincerely admire such consistent and dedicated service; it's not something I would care to have to do for any employer.

Perhaps James likes the back-to-basics approach at the distillery, for this is a very hands-on operation with few computerised controls or systems. Today, some distilleries are so mechanised that they could, in theory, be controlled by one man or woman from a laptop computer in the comfort of their own home, or the more appropriately corporate setting of a frequent flyer lounge in some distant airport. Some distilleries feel as though they are.

Nothing of the kind is to be found here. The distillery is small, with just one pair of stills, though in the process of enlargement as I write. It seems, therefore, to be now in calmer waters; happier times than previously, no doubt a beneficiary of the boom in global whisky sales. James told me that the output was now fully reserved for sale as Arran's single malt, which is impressive, and the whisky has a growing band of enthusiastic followers, many of whom have purchased their own cask of whisky. Arran raised a significant proportion of its original funding in this way and is one of the few distilleries that remains prepared to sell a single cask to the public. Most cannot be bothered with the paperwork, or the persistent enquiries of the owners, and while I quite understand this, I can't help feeling that something important has been lost.

Gone, sadly, are the days when the local laird would order a cask or three for his cellar. Gone, the opportunity for a nearby pub or hotel to offer a unique taste of 'their' distillery. Gone, blown apart by the winds of global commerce and dispensed with in the name of efficiency, is one small, emotional link between distiller and local community. The spirit of whisky weeps unnoticed. Well done, Arran, for maintaining this tradition, even if a majority of the private casks will eventually make their way off the island.

Gone, too, is one of the distillery's most interesting releases.

In 2004 Arran collaborated with the Agronomy Institute at Orkney College who were seeking to revive the production of bere, an ancient strain of barley. Arguably, this was the basis of all Scotch whisky production from the earliest periods through to the end of the nineteenth century, when modern plant breeding introduced new and higher-yielding hybrid barley varieties.

You will recall, of course, that Robert Burns celebrates it in *Scotch Drink*, though he spelt it 'bear'.

> *Let other poets raise a fracas*
> *'Bout vines, an' wines, an' drucken Bacchus,*
> *An' crabbit names an'stories wrack us,*
> *An' grate our lug:*
> *I sing the juice Scotch bear can mak us,*
> *In glass or jug.*

One single batch of Burns' 'bear' was distilled at Arran, which in 2004 produced just 4,890 bottles of 'Bere Barley Bottling', albeit that these were bottled at 56.2% abv.[3] It was an experiment ahead of its time. Had Arran continued with the use of bere the distillery would now have something of very great appeal to the connoisseur market. As it is, only Bruichladdich persisted with this noble, ancient grain and we shall learn more of their work in due course.

What then, of today's whiskies? There are a number—I might in an uncharitable moment suggest too great a number—of special releases, though the frantic pace of these has slowed in recent years, and a sensibly restrained core range. This comprises single malts at ten, fourteen and eighteen years; the Machrie Moor peated expression at two strengths (46% and 58.2%) and a number

---

3 'abv' = 'alcohol by volume', a measure of the strength of the spirit. Whisky is normally bottled at 40% or 43% abv, thus this particular Barley Bottling was unusually strong.

of cask finishes. All, of course, are available at the distillery shop and through various online retailers, as this is not a whisky you are likely to find widely distributed on the high street.

Though I wish them well, this is not, I think, a whisky that would be greatly missed had it never existed. They are pleasant, easy-to-drink, but not especially memorable. The eighteen-year-old single malt seemed to me the pick of the bunch and, at £75–£80 a bottle, represents fair value in today's market.

I can't help feeling, however, that Arran has been more than fortunate in having a very patient long-term owner in Leslie Auchinloss, and that without his consistent funding would never have survived to flourish in the sunlit uplands of today's buoyant whisky market. I hope his investment pays the rich dividends it deserves.

It has been a long and, at times, tedious process to get the distillery to full production and the tourism side of the business has been hugely important. Never forget that a bottle sold through the distillery shop is not only the most profitable bottle any distiller will ever sell but emotionally builds a link between purchaser, brand and distillery, the significance of which cannot be understated. While this is true of every distillery, such sales evidently play an especially important role here, something a far-sighted consultant once observed to his client at the HIE.

Notwithstanding my less than enthusiastic comments on the whisky itself, I hope that Arran continues to prosper and that it will, during these years of plenty, build for itself a stable and secure base for the harder times that will surely follow. Evidently, the ambition is there, as demonstrated by the expansion of the Lochranza site and the new distillery to be constructed at Lagg.

It is to the company's credit, too, that it has remained independent for the short years of its eventful life. There have been takeover approaches, of course. However, I shall refrain from idle speculation on why these did not proceed and simply rejoice in the fact that Arran remains one of the few distilleries in Scotland that is not ultimately controlled by one or other of the giant multinationals who dominate the industry.

Is that an essentially emotional reaction? Well, yes, it is. It would be foolish and perverse to deny the contribution made by the larger concerns. They have invested in the latest and best plant and equipment; their portfolio strength gives them the leverage in the market necessary to nurture their smaller brands; they have improved health and safety, and they train and pay their people well.

For all that, they are distant and remote and inclined to trade their assets like so many bags of sugar. I note, however, that Mr Auchincloss lives on Guernsey, though originally a native of Kilmarnock. It will not have escaped your notice that that was once a town where whisky ruled, until the Johnnie Walker blending facilities were eventually closed.

But I digress, and we must leave Arran and move on to Jura, an island entirely different in character and personality. This we achieved by means of the car ferry from Lochranza; a landing in Claonaig (observing there the bus shelter); a drive across Kintyre; another, longer car ferry from West Loch Tarbert; some sustaining Caledonian MacBrayne chips; a drive across Islay from Port Ellen to Port Askaig; yet another ferry and — on Jura at last — yet another bus shelter. No one said island hopping would be particularly easy or convenient.

I mentioned earlier a rather fine pen: a distinctively English Yard-o-Led fountain pen, which for this chapter I filled with Visconti's standard dark blue ink. Many years ago, it was a generous gift from the late Iain Russell, founder of the Arran Provisions business, with whom I spent many 'interesting' hours on Arran and elsewhere, discussing plans for his fledgling Arran Aromatics venture. The barrel is elaborately worked in a pattern reminiscent of the guilloche engraving on a fine watch face and is reassuringly heavy in the hand. I have allowed the silver to tarnish as it is exposed to the air; it feels as if polishing it would somehow cheapen the effect of age. The slightly faded patina reminded me of the Arran of old and somewhat of Iain's roguish charm.

Jura—a red island, I think, more of deer than people, and one where we shall meet both whisky and gin, along with a strange and varied cast of characters including, it is alleged, a werewolf.

3

Jura

Burning money

I HAVE DECIDED THAT JURA IS A VERY RED ISLAND — I
don't know precisely why, but it feels very red to me and so a very
special effort is required, perhaps to reflect the sheer awkwardness
of getting here. As I keep telling people, it's easier, quicker and
a great deal cheaper to fly to New York (unless you live on Islay,
obviously, when all that's required is to hop onto the small ferry
from Port Askaig).

In consequence, this chapter simply cries out to be written in
red ink. I'd really like it to be printed in red ink, but that would
look a little odd and I don't think the publisher would agree, but
try if you will to imagine these words in red. Now you may think
red ink is just red ink, a negative balance on an accounting sheet
somewhere or the kind of red ink that schoolmasters once used to
mark your work, but you would be sadly in error.

So, what to choose? Waterman makes nice ink, but then so does Visconti (the dashing Italian, in its handsome bottle) and Monteverde and Diamine, which has been made in Liverpool for nearly 100 years and comes in some 80 different colours, and good old Parker Quink of schoolboy memory and so on.

But I've gone to Japan, to the top of the ink tree with Pilot's Iroshizuku Momiji shade, intended, according to the manufacturer, to evoke 'the bright red leaves that are iconic of a Japanese autumn'. However, not to be perverse, I can't quite agree.

No, the Momiji reminds me more of the faded red of an elderly Cardinal's clandestine galero with its elegiac air of mortality, offering such a vivid and poignant contrast to the jaunty tone set by the scarlet biretta that the monsignor would have worn in life. As the galero first fades and then crumbles to dust so the onlooker is reminded of the passing of earthly glory—not that this is an association one would expect to come easily to the mind of a Japanese ink manufacturer.

However, moving from ink to gin, let us encounter Jura gin. An exhaustive and enjoyable tour of the Jura distillery had been completed and we were happily eavesdropping on the conversation of two young and extremely earnest backpackers (yes, Americans) when a passing stranger asked me, apparently at random, if I was interested in the distillation of gin.

This startling yet strangely perceptive query was addressed to me in the Antlers Café, which is just along from the distillery on the main street of Craighouse, Jura's principal settlement. It is not a substantial or overly formal establishment but I would thoroughly recommend it for plain comfort, good value and wholesome food. I have eaten there more than once and, not expecting Michelin excellence, I have not been disappointed. But I would be disappointed if you did not care for it.

Apparently, some gin had recently been distilled on Jura. A

brief discussion revealed that it was not yet on the market. However, it had been trialled only the previous week on the happy locals, who it seems don't all drink whisky, and was shortly to go into production. If I called at the distillery perhaps some would be available to try.

An even briefer discussion revealed that the would-be distillery was to be found somewhere at the Ardlussa Estate, at the north end of the island, my informant being a trifle vague as to the exact location. This was unfortunate as Craighouse is an inconvenient eighteen miles or so distant from where the gin was thought to be located. Ardlussa is therefore nearly thirty miles from the ferry, along a less than fast single-track road. George Orwell, who lived here for two years from April 1946 in an attempt to get some peace and quiet to write his greatest work, 1984, described this as 'the most unget-at-able place'.

He wasn't wrong: we could either attempt to get at the distillery (and assume that someone was there and amenable to granting a spontaneous tour to a random stranger claiming to be writing a book) or get the last ferry back to Islay, but not both. After the very briefest of discussions, the ferry won. I was not destined to see the making of Lussa Gin.

But as this trend of the micro-distilling of gin is gathering pace almost everywhere across the UK, I was intrigued to learn more, and further enquiry established something of its story. The production is on an understandably small scale, using a Portuguese still of 200 litres capacity (if you would like a point of comparison, the stills at Jura distillery are more than one hundred times larger). These little alembics have proved extremely popular amongst the growing community who aspire to become distillers: they are cheap, easily available, simple to operate and look good. If a venture is a success and demand follows, another still can be added quickly to boost production and the manufacturers offer them a range of convenient sizes.

The more well-known manufacturers of larger stills, serving the more established end of the industry, or those start-ups with patience and deep pockets are inclined, in private at least, to be slightly disparaging about their smaller rivals. Well, they might be, but I take the part of the novice. The budget Portuguese still allows you to get into the market quickly and economically and, if you fail fast, at least gives the consolation having failed cheaply. There are worse fates, especially as it may have some second-hand value to another hopeful.

The Lussa gin story is, like so many others, predicated on its location—which, it must be conceded, is unusual and distinctive. It's the creation of three local ladies, one of whom, Claire Fletcher, arrived as a 'wired' (her word) twenty-three-year-old as part of the video team filming the KLF (said to once have been a popular beat combo), of whom more later. The ladies are not just gin lovers, they say, but adventurers.

You'd need to be, living on Jura. Such a location is not for everyone. While Jura is an extreme example, island life can be demanding. I certainly wouldn't care for it. The weather would wear me down for one thing and the peculiar combination of isolation and yet exposure of your life to your few neighbours is a troubling prospect—or so it seems to me.

The Lussa ladies, though, value their connection to the land, the raw edge of their existence, the physicality of the wild and remote island and an intense relationship with the community that lives and works here. Presumably they are like-minded souls; I can't see how it would work otherwise. The Fletchers, who own Ardlussa estate, are said to be the only estate owners actually resident on Jura. Respect!

Andrew Ervin's novel *Burning Down George Orwell's House* is set on Jura, much of it between Craighouse and Ardlussa. Ervin presents a picture of an eccentric, booze-fuelled community on the edge of communal derangement, the events of his novel culminating in a drink-sodden hunt for a werewolf on the night of the summer solstice. It is presented as a time-honoured, traditional ritual, as is the systematic gulling of visitors to pay for copious quantities of whisky. It's probably best if I make it clear that it's a work of fiction. There are, so far as I know, no werewolves on Jura and a man I met there once bought me a drink. He may have been a local, and he certainly didn't have any hair on his palms. It's something I look out for.

Ervin is, at least, lyrical about the island's whisky, but his novel was published just as the earliest plans were being laid for Lussa Gin so sadly it doesn't make an appearance. However, the ladies may consider that their good fortune. According to the distillery, all the botanical ingredients are grown, gathered and distilled on Jura. There is, naturally, a website should you want to know more. You can purchase a bottle there, though at the time of writing, Lussa Gin has yet to make an appearance on many retail shelves. I hope they get a move on, because those shelves are getting very crowded and you need to be something special to even cram in at the back, let alone appear at the all-important eye level. In fairness, as we go to print it has been launched, apparently to some acclaim.

As there was insufficient time for us to make the trip to Ardlussa and back we went in search of otters. On a previous visit to Jura it had been common enough to see them from the road running through Craighouse, where a pair played happily for us in the surf and on the shoreline. No one gives them a second glance and the otters seem happy to ignore the few people.

On this occasion, there was some brief excitement when we thought we had spotted a young otter, but that turned out to be merely a large rat, which dashed our hopes. Apart from that, we walked to one end of the village and back to our car, passing the distillery's corporate house and entertaining facility, Jura Lodge. This is a most entertaining space.

If, for a moment, we accept that George Orwell's cottage, Barnhill, is the second most interesting building on Jura (Andrew Ervin's hero didn't burn it down, by the way) then Jura Lodge must surely be the third. We'll get to the most interesting in good time but Jura Lodge and I have history.

The building is a solid three-storey stone construction, white-washed in the Hebridean manner, stark and forbidding, standing to the right of the distillery as you look at it with the harbour to your back. Unfortunately, someone has erected a single-storey building housing the distillery's cooperage and a small space used for ceilidhs right in front of the house. This tends to diminish its impact. There is an elegant Georgian look about the architecture, which, given that it was built around 1810, suggests it was a fashionable addition to the village and would originally have been extremely striking when seen against its modest neighbours.

Alfred Barnard, writing around 1887, describes the whole building complex as looking 'more like a castle than a Distillery' and notes that what I take to be the Lodge is 'the highest building on the island, and the most elevated portion forms the summer residence of the proprietors'. As ever with Barnard, a certain caution is advisable; he was no stranger to hyperbole and, as he spent only one night on Jura and it seems unlikely he visited the grand estate houses, he may have taken the 'highest building' accolade on trust. Notwithstanding that, it will have been an impressive sight, especially when compared to the line of workmen's houses which then formed the main street of the village. It remains quite striking.

From the 1960s, the Lodge was used as accommodation for one of the distillery's management (I was told the engineer, though no one seemed quite certain) and, as was historically the case, the upper floor was reserved as an apartment for the then owner. Apparently, Donald McKinlay, whose company began operating Jura when it reopened in 1963, was a frequent visitor, enjoying family holidays here.

And why not? It's a favoured spot and there are unrivalled views of Craighouse Bay from the windows of the top floor. If I owned a distillery I would want such agreeable lodgings.

But time passed. Jura was eventually purchased by Invergordon Distillers in 1985, around the time I first visited and was shown the Lodge. It was then vacant, the engineer having moved out in favour of buying his own house. With the distillery in corporate hands there was no family to visit and bring the upper floor to life. Accordingly, it presented a forlorn appearance. It was then possible to rent one of the apartments as self-catering holiday accommodation, but the rooms were in need of urgent redecoration (in my memory at least) and my guide dolefully explained that visitors were few and far between. I did not find that hard to believe.

Invergordon was later bought by Whyte & Mackay and that company went through a bewilderingly rapid change of names and ultimate ownership. By 2001 it was owned by Robert and Vincent Tchenguiz and their partner, Vivian Imerman, a colourful South African entrepreneur who had made his fortune buying and subsequently selling the Asian interests of the Del Monte fruit company.

By 2005 he had acquired around 60% of the company and was eventually to receive some £380 million as his share when selling the business to another colourful entrepreneur, the Indian Dr Vijay Mallya. But he had overreached himself and was not to prosper. Having lost control of his United Spirits group to Diageo in something of a fire-sale, they quickly sold on Whyte & Mackay to its current owners for £430 million. Somewhere along the line, some £200 million of value had been destroyed in just seven years. Dr Mallya was last reported in dispute with the Indian Government and in exile in London with many of his personal and corporate assets seized and auctioned by the Government and his numerous creditors.

But before that, Vivian Imerman fell in love with Jura and saw the potential of the Lodge for high-end corporate entertaining and luxury rentals. Accordingly, he commissioned the highly improbably named Bambi Sloan (yes, there really is a person who answers to this), a Paris-based but American-born interior designer, to remodel the apartments, giving her a free hand and a lavish budget.

Ms Sloan's take on the hunting lodge design tradition is perhaps best described as eclectic. Though seldom seen by the Diurachs,[1] it has not escaped their attention and is regarded as a place of mystery and eccentric extravagance, as well it might be. Seldom seen officially I should add. 'Off the books' tours were not unknown, especially in the early days of the Lodge. But as you probably aren't allowed in, permit me to open the doors a little.

Amongst the remarkable features are a suit of armour painted in white gloss paint; a wall of antlers, some hung upside down (Bambi appears to like stuffed animal heads and antlers, which may be seen in a few of her interiors, though I always understood that the upside-down antler foretold bad luck and even ill fortune to the home); a 'music room' with a very large number of light fittings, the bulbs for which were—or so the frustrated housekeeper assured me—all but unobtainable at any cost in Scotland, hence bad luck for the housekeeper you may feel; a flamboyant, improbable and contrived homage to a leisured gentleman's cabinet of curiosities, a trophy perhaps of the Grand Tour, and very large bedrooms with

---

1 The natives of the island; there is no letter 'j' in Gaelic, where the island is known as Diura.

oversized bathtubs curiously located in the middle of the room.

Another striking feature was the wood-burning stoves that had been bolted shut on the orders of the distillery manager immediately after Ms Sloan departed the island, as he was all too conscious of the proximity of the chimney stacks to the warehouse roofs. Sparks and embers are not good companions for many thousands of casks of high-strength whisky.

The lodge was briefly promoted for luxury 'experience' rentals at a reported £2,500 per night, but before long the company was forced to abandon this plan, not through shortage of demand, but because of the difficulty of access via the one external stair and the consequent health and safety risk to guests in the event of fire. It simply wasn't 'up to code' and so was closed to paying guests.

Today it is used as a corporate facility and, on occasion, as an artist's retreat. Distinguished guests have included the authors Alexander McCall Smith, Val McDermid, Janice Galloway, Stuart MacBride and even Will Self. He seemed to me a curious choice, given his infamous use of Class A narcotics, but perhaps the tawdry frisson of danger and notoriety that once attached to his name appealed to the brand ('let's try for "edgy",' said some PR manager). Full disclosure: I was invited there myself for a few days some years ago, though not as part of the Jura Writers' Retreat programme, as presumably I'm not cool enough. Sadly, after flourishing briefly, that seems in abeyance.

Jura Lodge seems to me to have very little to do with Jura distillery and even less to do with its island home (but let's admit it was interesting to peek round its doors for what is probably the last time—I don't imagine they'll invite me again). It is an eccentric, arbitrary and wholly imposed confection, and maddingly engaging and amusing for all of that. I love it, but I couldn't live there.

It stands in contrast to its neighbour, the distillery, which continues to promote itself as 'established 1810'. It's a nice line:

there was a distillery here in 1810 which operated until 1901 when it was closed and dismantled. Eventually the landlord stripped off the roof to avoid property taxes and the whole place went to rack and ruin. It never distilled again.

Now in the manner of so much of Highland Scotland, Jura is divided into grand sporting estates. In the early 1960s two of the landed families got together with Charles Mackinlay & Co to establish a new distillery, hiring the renowned William Delmé-Evans, who had earlier built Tullibardine distillery in Perthshire, to design and build it for them. Essentially, this was a job creation scheme designed for the economic regeneration of the island, and the proprietors, practically enough, looked to fulfil a market need for fillings (i.e. whisky for blending purposes) as opposed to the new-fangled yet all-but-unknown single malt which was just beginning to enter the consciousness of the more discerning whisky drinker.

Thus, from the outset, Jura's whiskies were specifically designed to fit easily into mass-market blends, especially those of Scottish & Newcastle Breweries who then owned Mackinlay & Co. Shortly before he died, Delmé-Evans recalled it clearly enough. 'It was our intention,' he told Whisky Magazine, 'to produce a Highland malt differing from the typically peaty stuff last produced at the turn of the century. I therefore designed the stills to give spirit of a Highland character, and we ordered malt which was only lightly peated.'

Though Barnard describes the make as 'pure Highland malt', which in the late nineteenth century is highly suggestive of a peated style, he does not specify or even mention a peat store, so I've often wondered how Delmé-Evans knew what Jura had produced in its earlier flourishing. Note, of course, that he was ever so politely aiming a shaft at the neighbouring single malt of Islay. He continued to run the distillery until 1975, shortly after which it was expanded and, a decade on, acquired by Invergordon, where its modern history begins.

But let us just recall that the 1810 designation is more than a little disingenuous. Virtually nothing of the original distillery survives, save some walls to the filling store and a foundation stone on its main building. The existence of the Lodge, which is the principal remnant of the earlier establishment, does give some credence to the story, but the marketing of faux heritage is all too prevalent in whisky. Sometimes this really does strain credulity.

However, having got here, you probably want to go inside. Well, there are tours, of course, and a small visitor centre, now regrettably inadequate for the numbers of visitors seen in the summer. Eight to nine thousand people annually doesn't sound like many but on a busy day it will feel like quite enough.

Graham Logan had recently taken over as manager; so recently in fact that he had resorted to using his predecessor's business card while he waited for his own. 'Willie Cochrane' is thus scratched out on the card I took away and Graham's name inserted.

Not that Willie is forgotten. I spoke to him later and learned that he was retained by the distillery as a 'consultant' and a roving brand ambassador—this being the industry's posh name for salesman—of his own special bottling of a twenty-two-year-old single malt known as 'One for the Road'. Quite right too. In the currently fashionable manner, it's naturally a limited edition, so I asked Willie about the contemporary phenomenon of collecting whisky for 'investment'. He merely snorted, a man after my own mind.

'Get it out and get it drunk!' That's Willie Cochrane's view, ladies and gentlemen, the considered wisdom of nearly four decades in a wonderful industry: whisky is for drinking, for enjoying, for sharing with friends. He made it and he should know.

Willie, incidentally, is not a Diurach. He's a Glaswegian who came to Jura in 1977 and never left, spending the final sixteen years of his career as manager. If it works for you, Jura's evidently a hard place to leave, and not just because of the appalling road and ramshackle ferry.

Anyway, back to the present day. Having been expanded since the initial construction (capacity was doubled in 1978), the distillery is cramped and the tour inevitably something of a ramble, not that the enthusiastic visitors seem to mind. Ironically, on a more or less empty island of 142 square miles, the distillery was built on a very restricted site where space is at a premium. Further expansion is going to represent a challenge for the latest owners who, in the increasingly baroque world of corporate domination of Scotch whisky, turn out to be a Philippines-based brandy distiller called Emperador Inc. (they also own Harveys of Bristol, but best not tell your grannie), itself part of a larger combine known as Alliance Global Group Inc. If you saw that on a brass plate you'd think they were spooks, something straight out of a Fleming or le Carré spy novel.

Right now, Graham was happy to assure me, things are looking up. The distillery is on seven-day working and the 1.4 million litres of alcohol they will make this year are all reserved for sale as single malt. This is, of course, a complete reversal of the distillery's original purpose in a little over fifty years and is a really remarkable achievement. Today the malt whisky it makes has a growing reputation, greatly advanced from the inoffensive and bland malt originally destined largely for blending. Happily, some seventeen people are employed here; a lot from a population of fewer than two hundred.

If you came to the distillery completely fresh, and with no knowledge of its history, I believe you would be pleasantly impressed. This is due in no small measure to a major decision made in 1999 to re-rack (to empty all the casks of whisky into new oak barrels). Some 27,000 casks were then in stock, so this was a considerable physical undertaking, not to mention an expensive one.

The distillery, entirely against the wisdom then prevailing in Scotch, had increased production around 1978/79 when others were cutting back and consequently had good levels of stock when the market recovered. The challenge was to improve the quality, and re-racking into better barrels is whisky's time-honoured answer.

As we strolled through the building, chatting to Graham, I observed something on the malt mill. It's the ubiquitous Porteus design, a type seen so often that we frequent distillery visitors no longer really notice it. This particular example—it's the four-roller version, by the way—dates from the 1950s. The most probable date is apparently 1957.

What interested me particularly was a small plaque bearing the name Ronnie Lee, a title of 'engineer' and a telephone number. It struck me that I had never seen one of these before, and then it struck me that perhaps I had never looked. I resolved to look more closely in future and, as you shall learn, there was much to see and for me to learn.

For the moment though, on to the whisky. There are quite a number of different Jura bottlings to try, some from the distillery's core range and a good few one-off special editions. The one I'd suggest, ignoring the supermarket standards and the more expensive one-offs, is the sixteen-year-old Diurach's Own. It's what Willie Cochrane drinks when he's not On the Road and he assures me that it is genuinely popular with the islanders, though I presume that at least some them of have switched by now to Lussa Gin.

Back in October 2014 the distillery did release a bottle commemorating the island's most famous temporary resident, George Orwell, who wrote his magnum opus on Jura. But bottles of 1984, as it was imaginatively named, sold out quickly and now command some £650 each. A brave new world one might almost declare. There is a thirty-year-old version for considerably less money but with a significantly less elaborate finishing regime. All whiskies are equal, it would appear, but some whiskies are more equal than others.

I recall that the 1984 was launched at an extravagant themed party in a suitably dystopian setting in a dingy London railway archway, with costumed guards at the entrance and even a 'Room 101'. There was a lengthy queue to get in but, as that didn't appeal to me, I walked past some of my cowed colleagues and informed the guard that I was a senior member of Airstrip One's inner party and that if I was not immediately admitted he would be sent for re-education. He looked confused, but channelling my inner Kim Jong-il had worked and I was promptly ushered in, reflecting only later that I would have looked very foolish indeed had I been sent to the back of the queue.

Anyway, enough terrible mangling of Orwell. I don't intend inflicting any more tourism information on you—down the memory hole with it, I say—but picking up the theme of equality I must give passing attention to the system of land ownership on Jura.

At one time, and for around three hundred years, most of the island was owned by a succession of Campbells, bar a portion in the north which is held today by the Fletcher family, now of Lussa Gin fame. Most of the well-heeled proprietors are absentee landowners, visiting primarily for stalking and field sports. Amongst the notable owners are Master of the Horse of the Royal Household, the billionaire Samuel, 3rd Baron Vestey and, supposedly, Lord Astor, stepfather to Samantha Cameron, wife of the recent British Prime Minister.

The 18,700-odd acres of the Tarbert Estate are widely reported to belong to Lord Astor but the legal owner is Ginge Manor Estates Ltd of PMB 58, Nassau, Bahamas, and there is apparently no way of determining the beneficial owners. Coincidentally, however,

David and Samantha Cameron are said to have celebrated their June 1996 wedding in Ginge Manor House, Oxfordshire, the residence of William, 4th Viscount Astor, and you may draw your own conclusion from the fact that the Camerons also holiday on the Tarbert Estate.

Even more interesting for our present purpose is the ownership of the Ardfin Estate and Jura House. In 1938, Ardfin was bought by the Riley-Smith family (then owners of the John Smith's brewery business in Tadcaster, Yorkshire, but eventually sold to Courage in 1970) who, I have heard it said, were seeking a bolthole against the possibility of a Nazi invasion of Britain. This, I should add, though it may seem a practical if expensive precaution, is a far-fetched and scurrilous rumour and I entirely disassociate myself from it. It wouldn't have worked anyway, as even the Nazis would have eventually reached Jura as there are many things to shoot at there.

A later Riley-Smith was partially responsible for the construction of Jura distillery and for many years maintained the tradition of the private bottling of a cask of whisky for the Jura Hotel in Craighouse, opposite the distillery. Such bottles, under the old Stag label, appear occasionally at auction. Good luck with finding one though. Even more recently, yet another Riley-Smith went into publishing and launched and continues to run *Whisky Magazine* and other equally fine and perfectly splendid related ventures.

However, the estate, like the hotel, has now been sold and here things get very red indeed; that is red, as in the colour of burning money. Ardfin is a modest 11,500 acres or so, with ten miles of coastline, seven islands (one for every day of the week presumably), and a c-listed, sixteen-bedroom house, a trifle careworn the last time I was able to take the briefest of looks at it.

Today, all this is the property of one Greg Coffey, a one-time Australian hedge-fund manager; presumably a very successful one as his net worth was reported by the *Daily Mail* as £430 million when he 'retired' in October 2012. Whatever the true figure, he is spending it in style, and I believe that today we are obliged to rate the Jura House complex at the heart of Ardfin as Jura's most interesting building.

Having purchased the estate, which was on the market for a mere £3.5 million, Mr Coffey closed the gardens and sold off all the farm stock. There was some controversy over the closure of the gardens, which had been a minor tourist attraction, but they seemed pretty shabby when I last visited and well overdue for complete restoration and replanting. His plan, however, is more radical: Mr Coffey is creating an 18-hole, par 73, 6,982-yard golf course.

There will, however, be no clubhouse, no boutique hotel and no public access. The course, reputed to cost a staggering £55 million, is exclusively reserved to Mr Coffey and his guests. You can look down on part of it from the main road (the only road) and reflect that while all golfers are equal, etc. Apparently, sand for the course had to be shipped in from Ireland. That's the story, anyway.

Lots of people seemed to be busy working on some buildings as far as I could see and an awestruck local later assured me that so extensive was the project that Mr Coffey had built a temporary village for his construction team, and was happily buying any house that was offered for sale to provide future accommodation for his staff.

As you will have gathered from my brush with the Open, I am not a golfer, but I know enough about golf and Scotland to know that, traditionally at least, it was not a game reserved to the rich and famous and that the common man was as common a sight on the course as his laird. This particular link with the past, a commendable one as it seems to me, appears to have been lost on Mr Coffey and his links, though perhaps he will eventually relent and offer at least a few public days to his neighbours.

This brings me, happily enough, to the topic of burning money, as if £55 million on a private golf course was not expenditure enough. I speculated idly if it occurs to Mr Coffey to wonder who will maintain his golf course after he has gone, or will it turn into a colossal wreck, its lone and level Irish sands stretching far away into the sea mists. Greg Coffey, fund manager, as Ozymandias, king of kings.

The reference to burning money, if you haven't guessed by now, is to the infamous 'art event' during which, in the early hours of 23rd August 1994, the self-styled K Foundation burnt a million pounds in Bank of England £50 notes in the rundown boathouse on the Ardfin estate. In a curious turn of events, the boathouse is now being restored as a 'refreshment stop' on Mr Coffey's golf course. I don't know if there will be a plaque on the wall, or perhaps a video screen sharing the film of the burning on a never-ending loop. I do hope so, it would be the most perfect metaphor.

The K Foundation was the creation of Jimmy Cauty and Bill Drummond, who had enjoyed a series of hits as the pop musicians the KLF, and managed to make some money (quite a lot in fact). This seemed to have greatly troubled them and they turned to art and art criticism, presumably as a form of therapy to assuage their guilt in having made far too much money far too quickly. I should have their problems.

A series of well-publicised provocations followed. Such as awarding Rachel Whiteread the K Foundation £40,000 prize for Worst Artist (a mirror image of the Turner Prize which Whiteread had just won, but double its value) and threatening to burn the cash on the steps of the Tate Gallery if she declined to accept it. To cut a long and frankly tedious story short, eventually they drew £1 million from their bank in £50 notes, hired a private plane, flew to Jura, drank some whisky at the hotel and then took the money to the Ardfin boathouse and burned the lot. Well, not quite the lot; legend tells that at least one enterprising islander recovered around £1,500 in slightly singed notes for a bumper payday: handed in to the police, the K Foundation forbore to claim it and so one Diurach scooped the lot!

That is not, of course, the end of the story. As an 'Art Foundation' Cauty and Drummond had the event filmed and then ran a series of presentations and discussions round the country beginning, provocatively enough, in Jura on the first anniversary of the event. As you might expect it was not well received. As one member of the audience put it: 'I think you have abused Jura, not intentionally, but you have. I think it's an insult to the people of Jura … to come here and show us the film and the burning of a million quid. People on this island haven't even got £1,000 in their hand. It's insulting.'

I think we can probably agree. Twenty years on it appears little more than a monstrously self-indulgent piece of egotism by two over-privileged dilettantes. To add insult to injury a book documenting the whole thing was subsequently published, containing a lot of grainy pictures of the money burning and some pretentious commentary to the effect that burning a million pounds was 'accessible in a way that fervently and naturally sidesteps current cultural quagmires of irony, style and postmodernism; it doesn't need ghettoised art genres to sustain or quantify it.'

I don't suppose anyone, least of all its author, has the faintest idea what that means, but perhaps the distillery could release a K Foundation bottling and charge £50 for it. Or just one bottle at £1 million. The owners, Whyte & Mackay, seem partial to that kind of thing.

I think it's a tribute to the good people of Jura that Cauty, Drummond and their entourage weren't immediately lynched. I'm sure there are quite a few places where they would have been lucky to have emerged with only broken limbs. By April 2004, Bill

Drummond was reported to have regrets over burning the money. Imagine that.

But it would be a shame to leave Jura on this bizarre and improbable note. I haven't really written about George Orwell, or the Paps, or the Corryvreckan, or the golden eagles (didn't see any), or the caves, or the departure of the last Campbell laird in 1938, or the ancient ruins, or even as much as I should about Lussa gin but I've tried to give an impressionistic view of the island and my reacquaintance with it.

With its dramatic contrasts—Jura Lodge and the Antlers Café are within 100 yards or so of each other, but several hundred thousand pounds distant—Jura is quite different in character from any of the other islands we visit on this ramble. But if you cannot visit, try to obtain a copy of *Jura: Taste Island Life* or its later companion *Spirit of Jura*.

These handsome little books seem to be the sole legacy of Jura distillery's Writer's Retreat programme, which appears to have last flourished around 2009. Such are the vagaries and fickle nature of corporate patronage. But they're attractive wee volumes, lavishly illustrated with wonderful photography, and with short essays and poems contributed by, amongst others, Alexander McCall Smith, Liz Lochhead and Sir Bernard Crick. Copies may be found for as little as £2.80 (post and packaging included)—a bargain when I tell you that the foot-passenger return fare on the Port Askaig to Feolin ferry (a five-minute journey) is £3.20.

And it was on the Feolin ferry, pushing stoutly against the fierce tidal flow of the Sound of Islay, that we sailed to Islay. But we did not linger; home was calling and our return was to be another day.

4

Mull

*Trying, not awfully hard,
to buy a distillery*

HAVING ONCE OFFERED TO BUY THE TOBERMORY DISTILLERY, it occurred to me that I really ought to visit it.

Strangely though, my visit to Tobermory, principal town of the island of Mull in the Inner Hebrides, was shortly to lead me to a grimy industrial estate, just off the M62, about ten miles outside Hull, where I met Ronnie Lee, one of the great unsung characters of whisky.

Who is Ronnie Lee? Well, I'd seen his name discreetly engraved on a plaque on the side of Jura's malt mill and then again at Tobermory, so that's what I wanted to know. Not unsurprisingly, my guide could only guess. It wasn't, in fact, until several weeks had passed and I encountered the self-same plaque on some Speyside distilleries that I was able to track him down.

'Who is Ronnie Lee?' I asked some more whisky people,

production die-hards, as it happened. 'Ronnie Lee,' they answered, 'you must know Ronnie Lee.' Well, it turns out that, in a world of tattooed, bearded Brand Ambassadors with their hipster tweed caps, and rock-star Distillery Managers who jet from whisky show to whisky show, Ronnie Lee is a quiet hero, for he is the man who mends the mills.

At Tobermory, as in so many other distilleries, the guide paused to show us the malt mill. It's a typical enough Porteus mill, painted in that distinctive, unshowy shade of dark red; sturdy; planted four-square in the mill room, ready to receive another load of malt, a quiet occupant of an unobtrusive corner of the distillery that just gets on with its job in a modest and understated way. A Porteus mill would never shout or draw attention to itself, you feel; happy to do an honest day's work and then await the next consignment of malt to be turned to grist.

As is normal—you must surely have had this experience—the guide explained that the mill was quite old and that the Porteus firm had gone out of business, outlived by the durability and simple excellence of a product so good and so enduring that it broke the company. It's not in actual fact exactly what happened but, as whisky lore rarely let facts get in the way of a good story, we'll gloss over that and move on.

I found out a little more about Ronnie and it turned out that he is a freelance millwright, spending most of his working life in maintaining, servicing and, an occasion, restoring Porteus and Boby mills for the brewing and distilling industry. Based in that industrial unit just off the M62, Ronnie works in surroundings as far removed from whisky's new image of sophistication, glamour and luxury as you might conceivably imagine, yet carries out vital work. As one of an elite group of engineers who keep these vintage machines in working order he is vital to the smooth and effective running of the industry.

Tobermory, which is part of Burn Stewart Distillers, is far from his only Scottish client. He works for Diageo, Chivas Brothers, Glenmorangie, Highland Distillers and Ian MacLeod Distillers, as well as a number of independent breweries and distilleries as far flung as Sweden, Italy and even the USA. If you want a vintage malt mill, or need one restored, Ronnie Lee is the man you call.

When I met him, he made a dramatic arrival in his fabulous red Aston Martin and regaled me with stories of his weightlifting days. It turned out he had been a Welsh Champion and a real contender for a Commonwealth Games medal.

You don't hear a lot about Ronnie Lee, which is why I've made this diversion to tell his story. He doesn't give master classes, doesn't blog and he's definitely not part of the fashionable crowd that increasingly follow whisky these days. Yet he's one of the most interesting people in whisky that I have met in a very long time and he plays a distinctive and individual role in keeping whisky running from the stills. He is, you might say, a unique cog in the whisky wheel. So that was something I learned from visiting Tobermory. But even before stepping inside the building I remembered something else that I learned between 2002 and 2005.

If you have pre-school children you've almost certainly seen Tobermory, even if you have never set foot in the place. That is because it was the setting for the 254 episodes of the BBC's *Balamory* series.

Now I can't say that I was a fan but I was aware of Balamory/Tobermory and the famous painted houses. Even if I hadn't registered the TV series, Tobermory features on a thousand Scottish postcards and tourist brochures, so I thought that I was quite prepared for the visual impact of its scenic harbour frontage. Here's the thing, though: remarkably, and quite possibly uniquely, Tobermory looks better in real life than it does in pictures. I was shocked, I can tell you.

The distillery, too, doesn't disappoint. It's crammed into one end of the main street, hemmed in by other buildings and jammed at the rear against a steep, sloping cliff. In that respect at least it reminded me of Oban. The site is now so constrained that the distillery can never expand; what you see is what they put here in 1798, when it was known as Ledaig. The original name translates from the Gaelic as 'safe haven' and the harbour has long been renowned as a fine anchorage.

Tobermory itself only really dates from a decade prior to the founding of the distillery when the town was established around a small farming community in order to promote the fishing industry. That was an initiative of the splendidly named British Society for Promoting the Fisheries of Scotland and Improving the Coasts of the Kingdom. In 1786 they had sent their surveyor, John Knox, around the west coast of Scotland with the eventual aim of establishing fishing communities to exploit the plentiful shoals of fish to be found there.

Knox took to the work with considerable enthusiasm, eventually recommending that forty fishing villages be built, with accommodation for fishermen and associated traders, school teachers and even a publican. The Society took a more hard-headed view and eventually built just four fishing stations,[1] of which Tobermory was one.

The appeal lay mainly in the potential of Tobermory Bay, not for the possibility of dredging for the Spanish gold reputedly to be found in the wreck of the *Florida*, a galleon of the Armada which blew up and sank in the bay in 1588, but for the apparently surer harvest of silver darlings, as herring came to be known.

Knox immediately recognised it as 'one of the best natural harbours in Great Britain', going so far as to suggest that it would

---

1 The other stations were at Ullapool, Lochbay (Skye) and Pulteneytown, part of Wick.

be suitable for a naval dockyard which could be 'employed by government as an arsenal for the facilitating of naval equipments and military embarkation in Time of War for America and the West Indies'. Presumably the American War of Independence, which had concluded in 1783 with the Treaty of Paris (we lost, by the way), was much in his mind. Talk about locking the stable door after the horse has bolted. Nothing appears to have come of this plan and indeed there was never much of a fishing industry here anyway; the fish were simply too far away.

But construction did start in 1788 on properties for the new occupants, and within a few years Tobermory was thriving. One particularly active and entrepreneurially minded tenant was a certain John Sinclair who, in April 1797, applied to the Society for the lease of some vacant ground at Ledaig. It seems that he had a distillery in mind, though his application was vague as to his precise intentions, doubtless because distilling had actually been prohibited by the government to preserve grain during the war with France after a crop failure.

As an aside for the curious, this suspension of distilling is recorded in a chapbook of 1796, *Cheap Whisky: A familiar epistle to Mr Pitt on the recommencement of distilling in Scotland* (a chapbook was a small, pocket-sized pamphlet sold by chapmen, or itinerant pedlars), which celebrates the resumption of distilling. As the preamble remarks: 'An unusual portion of joy has been thereby diffused among the lower classes in Scotland, who indulge the pleasing hope of again tasting their favourite beverage, the high price of which, had almost amounted to a total prohibition.'

Permission was granted for a brewery, but he appealed and in 1798 built the distillery. It must have been a frustrating time, for there were further suspensions of distilling the following year, also recorded in popular culture through the composition in 1799 by the renowned Perthshire fiddler Neil Gow in the mournful lament *A Farewell to Whisky*. Fortunately, two years later he was able to compose the altogether jauntier strathspey *Whisky Welcome Back Again*. Both remain a vibrant part of the folk repertoire.

Understandably then, the distillery seems to have been a modest enterprise initially, producing a mere 292 gallons of spirit in the first year, but Sinclair's various ventures were successful enough and he was able to retire at the age of forty-three in 1813, devoting himself to building up a large estate and maintaining his extensive family and considerable local charitable interests, having presumably appointed a manager at the distillery.

The distillery's own website is disappointingly vague on the history between the late eighteen century and the 1970s. We know that Sinclair withdrew completely from the distilling business around 1825 and the distillery closed in 1837, when he was still listed as the licence holder. It does not appear to have worked again until 1878 but was flourishing in 1885 when Alfred Barnard visited. Like Knox before him, Barnard was struck by the favourable position of the harbour but lavishes more attention on the water source and florid descriptions of the flora. His audience were as interested, it seems, in the descriptions of his journey and the scenery he passed through as the facts and figures of the distilleries themselves; his tasting notes, if they occur at all, could generously be described as scanty. In this, he is representative of his age, viewing Scotland through the romantic prism of the novels of Sir Walter Scott and Landseer's scenes of Highland grandeur.

Unusually for Barnard there are actually two illustrations accompanying his short essay but, as so often with this most frustrating of commentators, only one of the distillery, and that is of the buildings seen from the harbour. The view is recognisable today, though, and his description of the distillery remains a broadly reliable guide. He records an output of 62,000 gallons of 'Mull Whisky' from the 'Old Pot Stills' of 2,530 gallons (wash still) and 1,710 gallons (spirit still) respectively. At the date of Barnard's visit, the owners appear to have been Mackill Bros, also noted as cattle breeders and farmers in the immediate neighbourhood. Twenty staff were employed.

They, however, sold out to John Hopkins & Company in 1890, who in turn joined the DCL (the Distillers Company Limited, the forerunner of today's Diageo which at this time, in 1916, operated as a group of quasi-independent competing companies), though the distillery continued to operate under the Hopkins name. In the little reception centre there is a photograph dating from the 1920s (the exact date is unclear) in which the legend 'Visitors may see over the distillery [on/by] Applying to the Manager' appears. This is painted across a doorway, so it may be that Tobermory was an early pioneer in distillery tourism, welcoming visitors at a time when few would have been open to the general public.

However, tourist visitors notwithstanding, the DCL closed Tobermory in 1930 and that would seem to be that.

The End.

However, as things turned out, it wasn't. In 1972 the distillery was sold to a Liverpool consortium, soon joined by the Spanish sherry producers Domecq. They greatly expanded capacity, doubling the size of the still house, but unfortunately the business collapsed just three years later. The likelihood is that this was due to over-production and the fact that the company had only one main customer.

A small amount of whisky produced at this time still survives and is bottled today in limited editions, priced from £2,500 to £3,500 depending on the exact age and expression. As they have probably sold all the bottles by now, I believe I'm safe in expressing an opinion which, even if it offends slightly, will not have any adverse commercial impact. You will immediately have deduced that I'm not a great fan: I've been lucky enough to try both these whiskies, and in my opinion they are over-aged and excessively dominated by wood flavours. Some may care for this but I find them too assertive, and not in a good way. The problem, I suspect, is that they weren't terribly well-made whiskies in the first place, a point to which I'll return shortly.

After the failure of the 1972 reincarnation the distillery lay silent until it was purchased from the receivers in 1979 by the Yorkshire-based Kirkleavington Property Company. A small quantity of spirit was distilled but there was no serious or sustained effort to market the product other than the confusingly named Tobermory blend, aimed primarily at visitors to the island. The whisky did not enjoy a high reputation in the trade, author Neil Wilson commenting diplomatically in 2001 that the 'malt had never gained the steadfast reputation within the industry which was and is still enjoyed by the other island malts'.

Production at this time was extremely modest and the owners' real interests and intentions were fully apparent when, in the early 1980s, the handsome warehouse building opposite the distillery was converted to residential occupancy as a block of apartments. This, I feel, was unfortunate. Not only do the ground-floor apartments face onto the main road into the town with little view other than the end of the still house, it means that Tobermory's production now has to leave the island as soon as it has been distilled. In the current market there would undoubtedly be a premium paid for an island single malt distilled and matured on said island, but this opportunity is no longer open to the present owners.

That could so easily have been Glenmorangie. The Kirkleavington Property Co. had closed the distillery in 1985 and made the staff redundant. It was apparent that they had no real plan for its future and it was rumoured in the trade that they were interested in selling, though made no very obvious move to do so.

By 1990 I was working as Group Marketing Director of the company then known as Macdonald & Muir Ltd, essentially the core of today's Glenmorangie plc. At that time, it was an independent company, quoted on the stock market, but largely controlled by the Macdonald family and related interests. The business consisted of the Glenmorangie and Glen Moray distilleries and single malts, some undistinguished blends and a number of contracts for supermarket own labels.

I had been employed to 'freshen up' the marketing effort and bring some new thinking and ideas to the business, a task which in hindsight I embraced with more zeal than tact. Having introduced one or two new products (including my claim to corporate fame, The Native Ross-shire Glenmorangie, first ever branded single cask, cask-strength release by any distiller), I took a lengthy look at the business. I didn't like what I saw.

The bulk of the company's volumes were then accounted for by a large number of undifferentiated secondary blends. Some had enjoyed success in the past, but their days of glory were long behind them; they lacked any support or any realistic prospect of support in the marketplace and, while they contributed something to our overhead, they returned very little in the way of profit. The company maintained a very dated and inefficient bottling line which even I could see was well past its design life and overdue for scrappage.

Our sole assets of any real unique distinction were the Glenmorangie and, to a lesser extent, Glen Moray distilleries and brands, though Glen Moray was employed, too often in my opinion, to prop up sales of the blended products or defend Glenmorangie from retailer price promotions. It occurred to me that we had one great skill, which was the making and selling of single malt whisky. It further occurred to me that everything else we did was a distraction and that the future lay in concentrating exclusively on single malt whiskies where, I argued, the market of the future lay for smaller companies unable to compete with the giant brands of our larger competitors. However, this piercing commercial acumen was not well received by my colleagues.

However, ignoring, or possibly not even understanding their scepticism and greatly seized with enthusiasm by this idea, I determined to see what sort of portfolio of single malt whiskies could be assembled. My researches led me to Tobermory, then silent. It seemed to me that, though the whisky was not of the first rank, there was considerable potential in the romance inherent in an island whisky, particularly where it enjoyed a unique status. So, surmising that the owners were indifferent to their property, I asked for enquiries to be made.

Thinking that a direct approach would only result in any price being immediately inflated, I instructed a solicitor to write to Kirkleavington's owner, one Stewart Jowett, on behalf of an anonymous mystery client expressing interest in a possible purchase.

And there the story ends. Mr Jowett never replied and, in the meantime, my MD had lost patience with my innovative approach and I was invited to consider my future. It seemed unlikely to lie in further corporate preferment and so an accommodation was reached, resulting in my departure from the business.

At the time, it was probably for the best. Though I can, and do, lay proud claim to my development of The Glenmorangie Native Ross-shire expression, it was neither appreciated nor understood by my erstwhile marketing team, who promptly withdrew it. They were an unimaginative bunch of corporate drones, however, and I felt vindicated by a generous and enthusiastic review of the product in The Independent by the late, great Michael Jackson, probably the most influential whisky writer of the present age.

The irony, of course, is that twenty-odd years on, the present-day Glenmorangie company have adopted precisely the strategy that I tried so naively and ineptly to introduce back in 1990. Their ownership of an abandoned and forlorn island distillery, Ardbeg in this case, has proved a spectacular success.

I was not so very far ahead of my time. Macdonald & Muir could have obtained Tobermory very cheaply. By June 1993 Mr Jowett was presumably reading his post once again and accepted an offer of just £600,000 for the distillery (plus £200,000 for the remaining stock) from Burn Stewart Distillers of East Kilbride, who now also own Deanston and Bunnahabhain on Islay.

Back then, Burn Stewart were a very different company to today. They worked very much at the value end of the whisky business, concentrating on cheap blends and private label bottling; exactly the market sector that I had wanted my employer to relinquish. The early production under their ownership was, therefore, unremarkable and Tobermory continued to make little or no impact on the single malt market which was, by then, beginning to gather pace.

Eventually, though, they concentrated more and more effort on their principal brand, the mid-market Scottish Leader and introduced some improvements and enhancements to Tobermory. The town's economy was also improving, with growth in the tourist trade and investment in the harbour facilities to attract leisure yachtsmen. Production was slowly increased at Tobermory and, importantly, the quality and consistency of the whisky began to improve and better packaging was introduced.

Burn Stewart were eventually bought by a Trinidadian con-glomerate, CL Financial Ltd, in December 2002, at which time the company was loss-making and the new owners' apparent financial expertise and other drinks industry interests appeared to offer an excellent future for the Scottish business. At first, all went well with the acquisition of Bunnahabhain distillery and the Black Bottle blend in 2003.

But, as so often in Tobermory's turbulent history, it proved to be a false dawn. The CL Financial parent became greatly over-extended in its other businesses, experienced what was euphemistically termed a 'liquidity crisis' and had to be rescued by the government of Trinidad and Tobago in 2009/10. Various assets were sold, including Burn Stewart, which fetched some £160 million when finally sold to the South African Distell Group. They are, at least, committed to the spirits business and, on the current evidence, appear to be financially stable and capable of the long-term investment and brand development that Burn Stewart so clearly need.

Up to this point in the story I've suggested that the output from Tobermory was less than exciting. Frankly, much of the whisky produced up to the late 1990s was pretty poor and some of the peated Ledaig expressions were worse. Most ended up in cheap blends, which was about all the spirit deserved.

Since then, however, there has been a steady improvement and the Tobermory single malt is now in my view considerably under-rated. Much of this improvement was the work of Burn Stewart's then Master Distiller and Head of Distilleries, Ian MacMillan, a veteran of some forty years in the industry.

Now working for Bladnoch distillery, Ian has a traditional view of Scotch whisky, maintaining that the best whisky is made in old-fashioned distilleries by experienced people, with the minimum of mechanised system or computer controls. Time spent in his company is bracing, for his views are forthright and rarely constrained by corporate guidelines. This is a man with a strong and committed point of view who is unafraid to express his opinions and back them up with the evidence of his whisky. Not everyone will agree with everything he says, but no one will be in any doubt as to where he stands!

Is he perhaps something of a zealot? Probably, but good for

him and for the distilleries that he controls. In my view, he, more than any other individual, has been responsible for the steady improvement in the quality of Tobermory's spirit, and its growing reputation. It's Burn Stewart's loss that he has moved on, but Bladnoch's great good fortune, and I anticipate a welcome increase in the quality of that distillery's release over the next few years.

On that note, perhaps we should step inside the distillery, noting, as we do, that much about the layout and surroundings would be familiar to founder John Sinclair and Victorian visitors such as Alfred Barnard. If your taste runs to older, slightly dilapidated buildings and you prefer your distilleries with the minimum of modern process equipment and computer controls then Tobermory will delight.

Though I visited on a bright and sunny day, and was accompanied by a bright and sunny guide, there was a pleasing crepuscular quality to the surroundings. The old stones of the distillery walls and the arrangement of the building around a courtyard have a severe, almost monastic air, and I would not have been entirely surprised to see a hooded figure, representative of some esoteric order of alchemists, flit silently through the gloom.

It's actually a minor surprise to me, even a disappointment, that Tobermory is not haunted. It's precisely the sort of distillery that should have a resident ghost—a tormented former owner, perhaps, or a disaffected employee slain in some tragic accident with a washback, whose spirit is forever fated to give body to the whisky. With so many eager tourists on the distillery's doorstep I'm disappointed that the marketing team aren't promoting ghost tours, but I make no charge for the idea.

Inside, all was silent as the distillery was undergoing its annual break for maintenance. Though expanded in 1972 and the maltings abandoned following the 1930 closure, much of the original may still be discerned, especially in the rambling and frankly inefficient layout. It is certainly not planned as a new-build distillery would be today, but with its many shortcomings it is all the more charming and engaging.

As you now know, there is a Ronnie Lee-maintained Porteus mill and four wooden washbacks dating from the 1970s. Everything along the way is quite cramped and awkward to manoeuvre through, but the still house is light and airy, thanks to a large picture window that looks out onto the main road into the town and across to the old warehouse, now converted into flats.

There are two pairs of stills, with current production around 800,000 litres annually, which I was assured is now wholly reserved for eventual sale as single malt. If pushed, output could be increased to something over 1 million litres a year, probably as much as this little distillery has ever produced in any one year of its long life.

Currently, two styles are offered. Tobermory is unpeated, using malt from Simpsons; whereas Port Ellen malt at 39 ppm of phenols is employed for the peated Ledaig style. Out of the two, I greatly prefer Tobermory, finding Ledaig a little crude and robust in style, with an excessive burnt rubber note. However, the peating levels have been reduced in recent years and the style may be getting a little more elegant and restrained. I suspect it will remain quite a mouthful for some time yet though.

I was especially partial to the Tobermory fifteen-year-old release, first available from 2008. This had been finished in Gonzalez Byass Oloroso sherry butts and represented exceptional value for money. Unfortunately, since I strongly recommended it in the first edition of my 101 *Whiskies to Try Before You Die*, it has sold out completely.

It's been replaced by a ten-year-old version, which is pleasant but not nearly as remarkable a dram as its older brother, supplies of which do turn up from time to time in the various online whisky auctions. Leaving aside the very expensive limited edition bottling of the last of the 1970s distillations, this is really the limit of available, officially bottled Tobermory, though Ledaig is available at ten and eighteen years and as a 1996 vintage, and there are, of course, a number of merchant bottlings also available.

Presumably, availability will increase in the future as maturing stocks are built up. If Distell aim to take the malt slightly up-market, as I hope they will, the fifteen-year-old Tobermory represents an excellent model and, dare I say it, is exactly the sort of whisky that I anticipated Tobermory producing when I made my furtive approach all those years ago.

Tobermory distillery, then, is a survivor. This may be due as much to its location as for any other reason. Had the property been located somewhere on Scotland's mainland, more central or accessible it would have been a great deal more valuable and would undoubtedly, like so many other urban distilleries, have been sold for redevelopment.

But there appears to have been no call in Tobermory for a supermarket, or a steakhouse, or more housing, and this, combined with the awkward and constrained layout of the buildings, seems to have ensured Tobermory's survival. 'A rising tide floats all boats,' or so they say, and the boating metaphor seems apt enough here. Tobermory has been the beneficiary of the rise in interest in single malt whiskies, and we are all the better for its survival.

It would be remiss not to include a few notes on the island and the town itself. Today it appears to thrive largely on tourism. Indeed, the canny traders on the main harbour frontage would seem to have reached an informal agreement of mutual non-competition for there looks to be little or no overlap in their offerings, albeit virtually all are aimed at the visitors' bulging wallets. Being a good tourist, I bought a handsome jersey there, and two tea towels. It seemed churlish to leave without some small souvenir, as the town has given itself over so completely to filling our previously unimagined demands for gifts and novelties. I certainly didn't realise that I needed tea towels until seeing them.

As for Mull itself, this was my first visit to what proved to be, on brief acquaintance, a completely charming and most attractive island, albeit with some challenging roads and determined local drivers. I can't say, and I really can't imagine, why I never visited here as a child. I blame my parents, but now I know it more than justifies a return visit.

Tobermory was exceptionally busy during our visit and so accommodation was provided, courtesy of the distillery, at the Bellachroy in Dervaig, some miles out of town. It dates to 1608 and claims the title of the oldest inn on Mull. While the architecture and layout reflect that, do try to drop by for a visit; the rooms are plain, but perfectly adequate for a short stay, and there is a pleasant bar with local ales and a decent if not extensive range of whiskies.

What did surprise me was the restaurant. There is a wholesome and well-priced dinner menu from which I ordered what I took to be gastro-pub style dishes accompanied by a pint of local, hand-pumped beer. It seemed appropriate to the place and the occasion so I was nonplussed when my choice was greeted with the slightest, but perfectly clear, supercilious glance from the waitress. I'd evidently committed some faux pas.

The answer became clear on studying the wine list, which features a surprisingly comprehensive list of clarets at up to and over £120 a bottle. Frankly, I was baffled to find them there and have been mulling this over ever since — this is one surprising island.

I finally departed from Mull with a question nagging in my mind. Over 100 years ago, in the late 1880s when whisky was booming, Tobermory was doing well. In fact, it was making a little over half as much again as nearby Talisker. Yet Talisker has gone on to considerable glory, fame and fortune even though, from 1916 to 1930, they were in shared ownership.

So, what happened? Why did Talisker boom as Tobermory faded? I made a mental note to investigate this conundrum once I got to Skye.

STOP PRESS: As I complete this book early in 2017, not one but two perfect illustrations occur of why the days of the conventional whisky book are limited. Firstly, the nice people at Burn Stewart have just sent me samples of two new whiskies they are about to release. The effect, of course, is to render my commentary of the various expressions from Tobermory not so much obsolete as frustratingly incomplete — I have been overtaken by events. This happens all the time to all whisky writers and that's why it's better for you to find information of this type on the web.

For the record, they were a delicious twenty-one-year-old Tobermory, finished in Manzanilla casks (mmm, spicy) and a nineteen-year-old Ledaig Marsala finish, which is quite possibly the nicest Ledaig I've tasted and seems to work surprisingly well. I don't think it's got to the 'elegant and restrained' stage yet but on reflection that is probably not what Ledaig is about. So, a punch in the face it is, though a well-bred 51% abv one. Who could take offence at that?

And secondly, I've just learned that the distillery was shut at the end of March 2017 for a two-year programme of upgrading and investment. But don't let that put you off, as the visitor centre remains open for your patronage and there's a lot to see on Mull, especially if you need some new tea towels and have the budget for vintage claret.

5

Islay

Dead crabs dancing

NORMALLY, IF YOU CAN'T FLY, getting to Islay involves a lengthy and frankly tedious drive to Kennacraig. The scenery might be fine but you can't look at it while driving as you will almost certainly plunge off the tortuous road and die, probably in great pain. This has no part in a successful road trip, so I suggest you pack some decent CDs, as you won't get a signal on the radio and only your passengers will be looking at the view. Apparently, it's nice. I wouldn't know: I'm generally driving.

Kennacraig is where you'll find the Caledonian MacBrayne ferry terminal, a grandiose and overblown term for a cluster of shabby Portakabins. After checking in, you wait in an interminable queue of caravans and motor homes; hope to meet someone you know but probably won't (they having had the good sense to stay at home); buy some vile sludge masquerading as coffee; throw this away in disgust; and eventually drive into the bowels of the ship where you abandon, if not hope, then certainly your car.

For amusement, you can stop on the way in Inveraray,[1] a town that has comprehensively sold its soul to tourism and now exists for no other apparent point than to sell knitwear, teas and coffees and provide toilet facilities to coach parties, and call in at Loch Fyne Whiskies. There, once upon a time, you could be cheerfully and imaginatively insulted by Richard Joynson, the former proprietor who possessed a unique approach to customer care. According to one highly regarded whisky writer, he used to be a fish.[2] This may explain much. Sadly, he has now retired and the shop has been sold to a national chain of whisky retailers. Richard is greatly missed.

Since I first went there, Islay has become incredibly fashionable in whisky circles. Some of us still view this with mild incredulity and some slight alarm. People (well, one person, Doug Johnstone, since you ask) write strangely violent novels about it, making Islay sound like a cross between a particularly inbred part of Arkansas and a lively Saturday night in downtown Baghdad.

It was ever thus. In 1777 the Reverend John McLeish of Kilchoman Parish observed that 'we have not an excise officer on the whole island. The quantity therefore, of whisky made here is very great and the evil that follows drinking to excess of this liquor is very visible on the island'. He regarded the island's inhabitants as 'wild barbarous people'.

Perhaps that's what the hordes of visitors expect. If so, they will be strangely disappointed, though the reputation of the female

---

1. At least you can if you're coming from anywhere other than Campbeltown. It would be extravagant, indeed almost perverse, to go to Kennacraig from Campbeltown by way of Inveraray.

2. According to Charles MacLean. He blames a typographical error — Joynson was actually once a fish-farmer, but sloppy editing omitted the all-important qualifier. At least, that's what they say now...

portion of the island for casual sexual encounters (so central to Johnstone's novel) is apparently well founded. Or so I'm told.

Perhaps it's genetic—an eighteenth century Dutch traveller, Johannes Hertz van Rentel, writing around the time of the Jacobite rising of 1745 on 'the excesses and intemperances of the People', observed that 'among the gentler sex of the lower orders of this island a wanton disregard for convention is often observed and many wallow in licentious abandon, freely coupling with any traveller. Entry to the Lists of Venus may be conveniently obtained for a measure of usquebagh, as they term the malt spirits distilled there in great quantities and to which the inhabitants are extraordinarily partial.' (From an unpublished manuscript, which I may not have transcribed entirely faithfully; something may have been lost in translation, or perhaps he was simply a randy old Dutchman. Or perhaps I made the whole thing up.)

Many visitors do indeed come today to track down some flighty birds. There are around fifty thousand geese to be found there, mainly white-fronted, pink-footed, greylag, brent and over three quarters of the world's barnacles (a sort of goose, not the gritty little shellfish that you find in rock pools and on the bottom of badly maintained boats). As well as the geese, you can apparently see choughs, corncrakes, hen harriers, linnets, woodcock and other small brown birds.

Not that I know the first thing about it. I just copied that list off the web. You'll be relieved to know that that's the end of the nature notes. If not relieved, then unfortunately you've bought the wrong book. You might as well hand it over to Oxfam now, because you're not going to enjoy the rest of it.

This remarkable avian assembly and the potential embrace of the island's womenfolk notwithstanding, most visitors are here—like me—for the drink. We're a little like Jig in Hemingway's *Hills Like White Elephants*—'That's all we do, isn't it', she says, 'look at things and try new drinks.' There you are: my life in considerably fewer than 140 characters.

Lots of visitors come from northern Europe where the assertively smoky taste of Islay whisky has found particular favour. Like curry fanatics searching for the ultimate vindaloo, these so-called 'smokeheads' or 'peat freaks' (labels they wear with pride) seek out ever more intensely smoky drams, argue about them obsessively and then award themselves extra points for drinking the fearsome stuff at cask strength.

Accordingly, a certain macho competitiveness has entered into the production of Islay whiskies, with the distillers attempting to outdo each other with the parts-per-million of phenols in their drams. So it's time. It's time to confront the 800-pound gorilla; to face up to the elephant in the room; to go mano a mano with—well, insert your own cliché here—the malts of Islay.

Islay, let's face it, is the definitive Scottish Whisky Island. There's a decent case to be made that it gets too much publicity, actually, but with eight operating distilleries and several more in the pipeline it's as hard to deny its dominant media presence as it is to ignore the pungent taste of most of its whiskies.

But here's my problem: everyone, and I do mean everyone, has written more than is healthy about Islay. From Alfred Barnard, who wrote one of his better pamphlets (actually commissioned, promotional literature, but let's not let that get in the way of the great man's reputation), to Michael Jackson and everyone in between and since, Islay has drawn whisky writers like drunks to a free bar.

Bloggers blog incessantly about it; the writing community have turned perfectly nice distillery managers into monstrous rock stars; the distilleries themselves (some of them anyway) feed the monster with a stream of provocative releases; and a great, seething mass of myths has been created that overshadows all other island whiskies and, indeed, looms large in Scotch as a whole.

The literature is intimidating. Neil Wilson and Jim Murray have written about Ardbeg, as has Gavin Smith. Marcel van Gils covers the *Legend of Laphroaig*, superfan Hans Offringa has waxed lyrical about the same distillery's double century, while German author Holger Dreyer managed a whole book on Port Ellen, which hasn't distilled a drop since 1983 and never will again.

The musician Robin Laing has committed the whisky legends of Islay to print while Bruichladdich had its own fan with a typewriter in the form of Stuart Rivan's panegyric *Whisky Dreams*. Such is Islay's magnetic attraction that at least two whisky writers have moved to live there on a permanent basis and a well-known third has an extensive plot of land on which he firmly maintains he will build a house. One day.

There are photo-essays, dedicated websites and, of course, the Islay Festival,[3] which draws visitors from around the world to a week of concentrated whisky madness. If you believe authors such as Doug Johnstone, Islay is quite the Wild West; his *Smokeheads* consume whisky on a heroic scale and the body count of a cocaine-fuelled weekend exceeds the days spent on the island. If you bought his version of the Islay story you'd imagine it to be a hotbed of corrupt constabulary, illicit distilling and gruesome violent death with an undertone of the ever-present possibility of some kinky sex. Little wonder that whisky writers want to live there.

It's a film pitch of a novel and, were it a whisky, I'd praise the nose, chew over the body (or bodies), but note that the finish fades away fast. Incredibly, it was published by Faber & Faber, once home to such literary greats as T.S. Eliot, Ted Hughes, Sylvia Plath, Seamus Heaney and … and, well, I won't go on as you've probably grasped my point: I'm simply jealous.

And I haven't even mentioned wine writer Andrew Jefford's

---

3 Otherwise known as the Fèis Ìle, the annual event starts on the last Friday in May.

Peat Smoke and Spirit, which seems to be everyone's reference point for the enchanted island. I'm afraid that I found it hard work: like eating a white pudding supper with a toothpick, it's a filling dish that's tricky to finish. Jefford seems earth-bound by his own ponderous solemnity, though it's only fair to observe that it's still in print after more than ten years so someone must like it. Again, I'm probably just jealous.

By now I've offended everyone that I've mentioned and, which is arguably worse, all the other writers that I've forgotten and haven't mentioned, but you see the problem: forests have died in praise of this island and its distilleries, so what can I possibly add that's fresh, new or different?

Actually, that's only the first problem. The second is this: everyone else has written in praise of Islay. They love the place, adore the whisky and envy the people who live there. I can't think of anyone who has a bad word for the so-called Queen of the Hebrides. Other than the obligatory beating dished out to Diageo for closing Port Ellen, there's little or no critical appraisal in the mass of adulatory and sycophantic copy.

I've been to Islay many, many times. I've come with the family and I've come on my own. According to my father I came here in the 1960s, though I can't say I remember very much about that and whisky certainly wasn't something that concerned me then, but over the last thirty years I've been a regular visitor.

I've come on a budget and on corporate jollies with no expense spared. I've come on the ferry, on the plane from Glasgow and in private jets drinking champagne (only Moët & Chandon, mind, nothing flash, but in general it's definitely the mèthode I'd choose).

So what's the second problem? Well, the fact is—H.M. Bateman moment coming up—peaty whiskies and me don't really get along. In truth, I find quite a lot of what's made here uncomfortably hard to drink. Being completely candid, I reckon one of the very best things off the stills here is probably The Botanist, and that's a gin.

To continue in this confessional mode, I've never really 'got' very heavily peated whiskies. I appreciate that isn't all that's made here, but that's what built the current frenzy of uncritical adulation and that's where Islay and I tend to part company.

Having said that, I can appreciate an Ardbeg, say, or a Lagavulin, or even an Octomore as a thing. I understand, intellectually, what it's about, how it was made, why it tastes as it does, but all too often these whiskies just seem like hard work. While I can drink one, or possibly two, in the interests of research and my own professional development, I can't honestly say that I greatly enjoy them and I'll happily put the glass aside in favour of something smoother or more mellow.

There, I've said it, and now we can get on. But while contemplating the Herculean nature of this part of my self-appointed task, I realised that I'd have to bring the big guns to bear: some serious writing had to be done here and that called for serious attention to the delivery system (as we writers almost never refer to our pen and ink).

The ink, as it happened, was easy: it had to be soft, warm and brown in colour in honour of the peat bogs that make Islay famous, and once again Pilot's Iroshizuku range came to my aid. Their Tsukushi shade is perfect.

'Iroshizuku' is a combination of the Japanese words 'iro', or colouring and 'shizuku', meaning droplet, the name embodying the very image of dripping water. Tsukushi is said to be inspired by the soft brown of a young horsetail awaiting the coming of spring, which is a pointer to the fact that this wasn't inspired by anything equine, but in fact refers to the horsetail fern. I liked it all the more when I discovered that: to me, this ink which flows so evenly and smoothly resembles the mosses and reeds that grow on Islay's peat bogs. If I filled my pen from the water draining off a freshly cut peat stack this is what I imagine the colour would be: an island captured and embodied in ink.

That was easy, and so too the selection of the correct fountain pen. Faced with a task of this significance, I turned to my favourite: a dramatic, yellow-bodied Visconti Van Gogh. Produced in limited quantities around five years ago, I had been saving this for a task of suitable magnitude. This pen, and this ink, are surely equal to the task, but will I sustain it?

Let's start with the time I threw my dinner in the sea.

It was at Bunnahabhain, the most far-flung of Islay's distilleries and the most inconvenient to reach. Before you jump to any unwarranted conclusions, I had not over-indulged and I was very far from inebriated. In fact, my mission was a merciful one, driven solely by parental concern and a desire for a quiet life; something all parents will recognise. It fell out like this.

It was 1987. I was working for the Glasgow blenders Robertson & Baxter (R&B). You won't have heard of them, but, take it from me, they were whisky royalty (today they are part of the Edrington Group). I didn't know then how very fortunate I was to be employed by them, a fact of which Mrs Buxton is wont to remind me from time to time.

Currently, Bunnahabhain is owned by Burn Stewart but, back then, it was part of the R&B stable. The original Manager's House had been left vacant when a modern house was built and, on occasion, it was possible to get the use of the older property for a very nominal rental fee. As the company then had a similar arrangement for their staff in two considerably more glamorous and luxurious villas in Spain there was never much demand for the accommodation on Islay, so I had reserved it for a fortnight's family holiday.

There is only one road in and out of Bunnahabhain, for the very sound reason that it was designed for access by sea. There is a sweeping stony bay here, just off the Sound of Islay and, in 1880, a team of labourers arrived to build a brand-new distillery for Highland Distillers Ltd.

Bunnahabhain was built from scratch in a previously uninhabited spot. It is not, it must be admitted, ever going to feature on shortbread tins. The building is stark and brutal. Barnard describes it as 'a fine pile of buildings...and quite enclosed', noting also 'a noble gateway', and Michael Jackson went so far as to compare it, not unfavourably, to a Bordeaux château. Well, on a sunny day perhaps, and the gateway is impressive, but I feel they must both have been in a generous mood. In other light, the buildings are gloomy and seem to loom over the roadway between the distillery and the sea. A little bit of maintenance wouldn't go amiss as the place has something of a careworn look.

Doubtless this is to do both with the situation and the materials used in construction. The original builders made free use of concrete and, with the buildings set tightly against the hillside behind, they present an unfortunately cramped appearance. In all honesty, the best view is had by looking out from the distillery to the pier, bay and across to the opposite shoreline and the spectacular Paps of Jura.

But for my first visit I was with my young family (two boys) and a seaside location where the children could play with the minimum of supervision, especially when our accommodation was for all practical purposes free, was ideal. It was easy to overlook the slightly intimidating and glowering presence of the distillery, especially as our house was to be found on reassuringly higher ground to one side of the main buildings.

One day I happened to be chatting with some fishermen who were landing lobsters and partens (brown crabs) from their creels. As it happened that I had several cartons of biscuits with me (full cartons, note, not packets) I proposed a swap, and so it came about that

I took possession of two fine crabs, which I intended for our dinner.

They were, of course, still very much alive, a fact which greatly impressed two young boys from the city and for an hour or so they enjoyed their new companions. The crabs were measured; the crabs raced; the crabs even danced.

I didn't think to ask the crabs what they made of this, as I had plans for them which eventually I explained to the children. As it turned out, I may have been just a little too explicit in my elucidation of the fate that awaited the hapless crustaceans. By the time I got to the part about the boiling water I was confronted by looks of some dismay: 'But that's going to hurt, Daddy,' came the comment.

Yes, friends, the crabs were saved. Under pressure from the family (for by now my wife had taken the beasts' side), and under cover of darkness, I was obliged to stumble across a stony beach and return our dinner to the sea, unharmed. I prayed that the fishermen did not see me, and we dined that night on shortbread.

The distillery was built at the start of the great Victorian construction boom, when whisky seemed to provide a licence to print money. Bruichladdich opened the same year (1881), but Bunnahabhain was fortunate in its ownership and was soon producing some 200,000 gallons of whisky annually. As a point of interest, as well as the date of their foundation, they have two other things in common: coal and concrete. This is not coincidental. Until this period, Islay distilleries would have run exclusively on local peat, but the advent of the shallow bottomed 'puffer' ships, which could easily serve coastal distilleries with coal for drying their malt, combined with the use of concrete in construction meant that Bunnahabhain and Bruichladdich could be built to serve the blending industry. To this day, both distilleries are best known for producing whiskies at the lighter end of the Islay scale of peatiness—no accident, you see; the reason for this lies in their shared history.

In fact, at this point, it's fascinating to rank the various Islay distilleries according to the output recorded by Barnard, as suggestive of their relative importance in the Victorian era. Here goes:

Ardbeg............. 250,000 gallons
Bowmore .... 200,000 gallons
Bunnahabhain....... 200,000 gallons
Caol Ila .... 147,000 gallons
Port Ellen.......... 140,000 gallons
Lochindaal .... 127,068 gallons
Bruichladdich .... 94,000 gallons
Lagavulin .... 75,000 gallons
Laphroaig .... 23,000 gallons

It's salutary to contemplate that league table for a moment, as it is illustrative of the vagaries of Islay's fortunes. Most obviously, two of the distilleries listed have closed. Both Port Ellen and Lochindaal (also known as Port Charlotte, after its location) have fallen victim to the decline in interest and demand for Islay whisky—Lochindaal during the round of closures that marked the 1920s and 1930s, and Port Ellen in 1983. Not so very long ago, if you think about it.

Also, note that comparatively recently both Ardbeg and Bruichladdich have been close to permanent shutdown and, at various times as recently as the 1980s and 1990s, most of the Islay distilleries have been on short-time working—the explanation, incidentally, for the shortage of really old Islay whisky and thus its exorbitant price.

In between Barnard's reports and the present day, there was the short-lived Malt Mill distillery at Lagavulin, which flowered briefly between 1908 and 1962. Most of the equipment was incorporated into Lagavulin, and the corporate entertaining suite there was created from the Malt Mill maltings, finally closed in 1974. Sadly, at least as far as anyone knows, Malt Mill was never bottled as a single malt, but the very last sample of the very last spirit run is displayed, looking curiously forlorn, in a wax-crusted bottle in a display case at Lagavulin.

It has, of course, been immortalised in the Ken Loach film *The Angel's Share*, where a mythical cask of Malt Mill is found and auctioned. Most people viewed it as an escapist and essentially light-hearted comedy, but Loach and writer Paul Laverty, who incidentally make Comrade Corbyn seem the life and soul of the party, insist that it deals with 'both practical and existential questions of the most profound nature'. Seen in that light, it's a dystopian vision with a deep moral vacuum at its heart: only crime allows the protagonist to escape his life of crime, and his sole answer to the dysfunctional society from which he sprang is to abandon friends and family, leaving those who paved the route to his flawed salvation to their no doubt inevitable bad end. It's all capitalism's fault.

However, whisky writer Charlie MacLean was very good in it, effectively playing himself with considerable élan and after I'd had a very stiff drink I did feel a little better about it. If you really want to see Malt Mill on film then try to find the Scottish TV documentary *Whisky Island* (1963), which discusses the problems of depopulation and the potential for tourism to regenerate the island's economy. It's online at the Moving Image Archive of the National Library of Scotland. It reminds us that economic development has been a long-standing problem, but if you look carefully, you'll briefly see some casks of Malt Mill being loaded onto a boat. Ken Loach was right about one thing: they would probably each be worth around £1 million today if one could be found. Probably more in today's febrile market.

Anyway, Malt Mill is long gone, but it isn't all bad news for Islay distilling: as recently as 2005, the small farm distillery at Kilchoman swelled the numbers to eight and, if the various prospectuses are to be believed, there are more on the way.

But what we've learned, apart from the remorseless and unforgiving nature of change, is that Bunnahabhain has declined in status since 1885. Then placed second equal, at least by volume, even its most passionate advocates would hesitate to rank it as highly today in public perception. Much of this is fashion, of course, but Bunnahabhain's relative obscurity is explained by the fact that for most of its life much of the output was required for blending. That is how it began life; it's why it was built and, until very recently, the spirit has been destined for blends such as Famous Grouse and Cutty Sark.

The previous owners made some sort of an attempt, not a very convincing one, to market Bunnahabhain as a single malt from the late 1980s, but it suffered from three problems: their heart was never in it; the name was all but unpronounceable and therefore off-putting and intimidating; and, back then, the spirit was unpeated. A peated whisky wasn't going to go very well in either the Grouse or Cutty Sark blends but, for the single malt market, peat was and is Islay's great calling card. So Bunnahabhain drifted along, not looking particularly comfortable as a single, especially when contrasted with its then stable mates of Highland Park and The Macallan. No one likes to be the awkward kid at the party, but poor Bunnahabhain really was the runt of this particular litter.

Its fortunes seemed to improve slightly when the distillery and the Black Bottle brand were sold to Burn Stewart for £10 million in 2003. But, as I have related elsewhere, Burn Stewart went through a period of instability and it is only comparatively recently that the impact of the new South African owners has been felt on both production and improved sales and marketing.

In the meantime, of course, the distillery's competitors have not been standing still, and Bunnahabhain does not enjoy the prestige or fashionability of some of the island's other single malts. The problem of the name won't ever go away, but it strikes me that the marketing team have almost perversely made life harder for themselves by adopting Gaelic names for the distillery's peated releases: Teiteach ('smoky'), Cruach-Mhòna ('peat stack') and Ceòbanach ('smoky malt'). I suppose it speaks to the distillery's origins and is intended to reflect heritage and provenance (two great words in the current marketing lexicon) but it doesn't make life any easier for the potential purchaser.

With a fair wind, though, Bunnahabhain could be a very big player. The overwhelming impression of the interior, once past the forbidding walls and the prison-like gates, is of scale. This is a result of a major 1963 expansion which took the theoretical capacity up to some 3.4 million litres annually.

Everything is on an appropriate scale: there is a substantial Porteus mill (another in the care of Ronnie Lee) and a fifteen-and-a-half ton capacity mash tun which the guide proudly assured us was the second largest in the industry; a curious distinction when you come to consider it, but impressive enough at the time. The six washbacks are similarly large, standing some five and a half metres tall, and the four stills are of Brobdignagian proportions, though rarely filled to capacity. That, and the significant degree of consequent copper contact in the distilling process, contributes to the relatively light style of the spirit — ideal for unobtrusive blending in fact.

If you do take the tour, you'll end up in warehouse nine, an atmospheric traditional dunnage space holding some 1,500 casks, where there are a range of interesting expressions to taste. These you can then purchase in a small shop created from one of the first-floor offices, which presents the charming impression of a time capsule. Nothing, you feel, could ever be rushed or hurried here amongst the calming atmosphere and wood-panelled corridors.

As the distillery was built on a green-field site, one of the first tasks for the builders was to create a village for the workforce. As there were some fifty to seventy employees at any one time, together with their families, they formed quite a community and, as Alfred Barnard records, as well as the 'neat villas' for the excise officers, extensive housing was provided along with a reading room and school where an elementary education was provided.

Most of the houses lie behind the distillery, but to the left of the main buildings as seen from the sea there is also a row of cottages. For some time, when these were no longer required by a reduced workforce, they were fitted out to be rented as holiday homes. Sadly, the cost of maintenance and constant upgrading proved too great and the cottages now lie cold and closed up, seemingly abandoned to the elements. The owners have apparently rejected a number of approaches to buy the properties, ostensibly on security grounds. It seems improbable that they will ever be required again for staff accommodation and, having also once stayed very comfortably in one of the cottages, I regret seeing their inevitable deterioration in the demanding Hebridean climate.

If you walk on down past the cottages, here's a tip: carry on round the headland and very soon you will see the dramatic wreck of the *Wyre Majestic*, a 338-ton trawler that is looking slightly less majestic since running aground on the rocks close to the shore in October 1974. Here's the thing: if you time your visit for low tide it's perfectly possible to hit the rusting hulk with a well-aimed stone thrown from the shore (there is no shortage of suitable missiles). It makes a very satisfactory noise and if you have small children with you, especially boys, they will be impressed by your manly skills. It's harder at high tide, but you're the adult and your adoring audience probably won't realise the advantage you've gained by timing your visit correctly. It fooled my kids anyway.

Finally, if you do ever visit, try to drop into the office where you will buy your tour ticket and say hello to Lillian 'the General' MacArthur. She's run the office there for close to forty years, watching owners and managers come and go. Some twenty years elapsed between two of my visits, but when on the second, totally unannounced and spontaneous occasion I dropped by she took just one look at me and, without missing a beat, said, 'Hello, Mr Buxton, and how have you been?' I think that deserves an honourable mention because I had, altogether rudely, completely forgotten who she was and to be greeted thus seemed almost supernatural.

To return to Port Askaig, unless you came by boat, which would be fun, it's necessary to return by the same twisting, awkward, single-track unclassified road that brought you to the distillery. The road ('as good as it was costly') was built by the promoters of the distillery for the greater convenience of operations, but was never designed for heavy traffic as barley and fuel came in, and whisky went out, by MacBrayne's steamer *Islay*.

Though the road offers exceptional views of Jura there was, until very recently, nothing else of consequence to see until the turning back onto the A846, Islay's principal road and the route to the Caol Ila distillery. Soon, however, there will be another stop to make, roughly halfway along the road.

This is the site of Islay's newest distillery, Ardnahoe, an initiative by the independent bottlers Hunter Laing & Co., a Glasgow firm established in 2013 following the break-up of the Douglas Laing & Co. business, also third-party bottlers.

When fully operational, Ardnahoe will have the potential to produce around 500,000 litres of spirit a year, though production will start at a more modest level and be increased to follow demand. There will, of course, be a visitor centre as part of the planned £8 million project. Planning permission has been granted and water rights obtained with the first spirit expected, as I write, in early 2018. I expect there will have been delays though; there usually are.

The name, incidentally, comes from the nearby loch. Hunter Laing's intention is to bottle and market the spirit as Islay single malt, their MD stating that 'we as a company can never get enough Islay whisky to satisfy our requirements'. There are apparently no plans whatsoever to make vodka or gin — 'we're a whisky company' is Stewart Laing's unequivocal view of white spirits.

In that, they will be at one with their production director, the highly experienced former Bowmore manager and Bruichladdich master distiller Jim McEwan, who is renowned amongst malt whisky consumers and who, amongst his other skills, has a genius for publicity and extravagant tasting notes. McEwan, now sixty-seven, had actually retired but was tempted back by the chance to design a distillery from scratch and shape the eventual style, which he describes as 'good whisky in good American Bourbon barrels for that traditional Islay flavour'. It'll be peated for sure. His appointment attracted considerable discussion on social media and has certainly increased Ardnahoe's visibility and credibility.

Full credit to Hunter Laing for making such progress on their project. Talk is cheap, but actually raising the funds, securing water rights and obtaining planning permission is altogether another matter, and hiring McEwan, even for a few years, is something of a coup. There are constant rumours and announcements of 'new' distilleries for Islay, but few seem to get further than the outline planning stage. I can think of at least three examples, though there may well be several more.

At Port Charlotte, the original Bruichladdich team under Mark Reynier announced confidently back in 2007 that they would be rebuilding the Lochindaal distillery, but that project is very much on long-term hold, with new owners Rémy Cointreau preferring to put their investment into production and warehousing at

Bruichladdich and supporting a greater global marketing effort.

Closer to Bowmore there has been an optimistic sign announcing the Gartbreck distillery since April 2014, but apart from the demolition of some old farm buildings to clear the site, very little seems to have happened. The owner, Jean Donnay, is already a successful and well-established distiller in Brittany and intends to create his Islay whisky using direct-fired stills.

However, despite repeated assurances that construction was about to start and that the first spirit would run in 2015, and then 2016, Gartbreck is believed to have run into funding problems and there is no apparent sign of any development on the ground. If it happens, Gartbreck will also feature visitor facilities, but will be small: the plans announced so far envisage no more than 55,000 litres of highly peated spirit annually. Early reports suggest that gin will be distilled, largely to support the initial cash-flow requirements.

Finally in this theatre of dreams, the well-known London whisky specialist, online retailer and confirmed whisky collector Sukhinder Singh of The Whisky Exchange is said to be the leading light in plans for the Farkin distillery just beyond the Port Ellen Primary school on the road towards Kildalton. Whisky folk will know this better as the road to the Laphroaig, Lagavulin and Ardbeg distilleries.

The Farkin Distillery Ltd was registered at a Campbeltown address in May 2016 with two active officers being John and Lesley Barford. Mr Barford gives his occupation as 'teacher', though with a date of birth stated as April 1953 he will surely be approaching retirement from the education community. Singh is tight-lipped about the project. Local rumour suggests that water supply may have become an issue—water for distilling is, surprisingly, a limited commodity on Islay despite the high West coast rainfall—but my enquiries elicited little or no further detail.

The contrasting fortunes of Ardnahoe and these three schemes illustrate just what a challenge it is to start a new distillery, let alone manage it through to the stage where mature whisky can be sold.

I'm asked often enough for my opinion by people who think they want to start a distillery, and my reply is always the same: go into a darkened room, I say, and lie down until the idea has gone away. It's good, honest advice and when I accidentally bought a derelict distillery at Tomdachoille, near Pitlochry, I took my own counsel—and I'm thoroughly glad that I did. Take it from me: it's a great deal easier to sit as a commentator, being wise after the event, than to actually risk your capital and many years of your life in starting a distillery, especially a whisky distillery. I've sold the property at Tomdachoille, by the way, so if you ever do see whisky with that name it's nothing whatsoever to do with me.

Having taken that little detour round Ardnahoe and its would-be peers, let's carry on down the road. The next logical port of call is Caol Ila, a giant Diageo distillery just outside Port Askaig. Incidentally, you may come across a Port Askaig single malt and be a trifle confused, as there is no Port Askaig distillery (certainly not a legal one). It's actually a merchant bottling by the London specialists The Whisky Exchange (Mr Singh again), and, as is normal with these third-party releases, to protect the distillery's trademark it has to be marketed under an alternative name: you can probably hazard a guess where it comes from.

When I was last at Caol Ila it was closed for some major main-tenance works so I spent much of the time in the manager's office enjoying an enormous seafood banquet which the distillery had laid on. There were scallops, mussels, prawns, oysters, dressed crab and a number of large and remarkably tasty lobsters, all washed down with some delightful chilled white wine. Consuming considerable quantities of exceptionally fine and very fresh sea-food is one of the unavoidable travails of corporate trips to Islay.

The occasion here was a Diageo-hosted tour of their facilities for a number of US-based writers and journalists and, as I write for one or two American trade titles, they were kind enough to invite me to join them, and an agreeable enough bunch they proved to be.

The office at Caol Ila must provide one of the most exceptional views from any office anywhere in the UK, as it has an entirely uninterrupted outlook through 180 degrees across the Sound of Islay to the Paps beyond. On this day the weather was spectacular and the visitors could hardly tear their eyes away from the magnificent scenery to take in the buffet. They seemed to manage somehow, however.

I spent a lot of the time looking for some living sea life. I've seen otters here on a previous visit and the distillery staff will tell you that it's not unusual to see whales, dolphins and porpoises. I did see something which may have been a porpoise (they do behave slightly differently to your actual dolphin and are very much smaller) but it can't go down as a definite sighting. Also absent were the golden eagles that are frequently observed over Jura.

If the wildlife wasn't on show to impress the group, the distillery did not disappoint, even though it wasn't actually in production. It can sometimes be easier to appreciate the distilling process in a silent distillery, as stills can be opened to inspect heating coils and an empty washback really brings home the scale of these giant vessels. Accordingly, I'm never completely disappointed to arrive and find the object of my quest is silent for one reason or another. On this occasion, the drama of the Caol Ila still room, which also enjoys a remarkable view to Jura, was enough to satisfy the most demanding whisky geek. Caol Ila isn't really set up to receive visitors, though it is open to the public and it is possible to arrange a tour. It's probably of specialist, minority interest, however, one for the completist.

I don't say that in any way to disparage or demean the whisky, though. It's actually one of my preferred Islay drams, particularly the older expressions, and it's probably more than a little underrated. This is presumably because the bulk of the output is reserved for Diageo's blends, especially the Johnnie Walker range. Caol Ila is a major contributor to the smoky Walker style, and with Walker's sales increasing round the world the blending team need all the whisky that's produced here.

There has been a distillery on this site since 1846, but very little remains of the original building, other than the Victorian warehouse that you encounter on approaching the distillery itself. This was rebuilt between 1972 and 1974 to a design by George Darge of the DCL's production arm, Scottish Malt Distillers Ltd (SMD), at the then remarkable cost of £1 million.

The engineering team were based at the SMD offices in Waterloo Street in Glasgow, from where a stream of similar distillery designs originated. SMD were great innovators, concentrating on improving efficiency and increasing production yields to serve their clients, the blenders in the various companies that made up the DCL.

SMD had been at the forefront of distillery expansion in the early 1960s, increasing the number of stills in their various plants by more than half, and a further burst of expansion occurred at the end of that decade into the 1970s. George Darge had been appointed chief architect around 1968 and was largely responsible for the introduction of the dramatic, fully glazed curtain walls seen on so many of the SMD distilleries of the period. For this, and other contributions to distillery architecture, he was elected FRIAS.

The glazed frontage had a number of functions: it allowed easy access for the replacement of the stills; it dissipated heat; and it attractively displayed the pot stills at the heart of the distillery. SMD's dominance of the Scotch whisky industry at this period is shown by the prevalence of the design: as well as Caol Ila, it may be seen at Teaninich, Brora, Linkwood, Aberfeldy, Royal Brackla (these latter two now owned by Bacardi) and others. Caol Ila, Darge's swansong before retirement in 1981, was said to be one of his favourites.

As fashions and tastes change, I do not think this particular, starkly modern design feature has aged well, and it tends to place the distillery very firmly at a particular point in whisky's history, not necessarily to its advantage. The rigid and uniform pattern of these designs and the associated engineering has subsequently been criticised as symptomatic of a period when whisky was being standardised and spirit character rendered more homogenised in what was perceived as the interests of the blenders who were then, of course, absolutely dominant in the larger companies.

Such a trend is perhaps inevitable when the overwhelming commercial imperative was the development of strong brands of blended whisky. It should be recalled that single malt sales accounted for less than 2% of the industry's sales then and that whisky as a category was coming under pressure in virtually all world markets from light white rums and vodka. Whatever their other merits these products are not noted for their forceful flavours. The consumer signalled a preference for blander alcohols; the industry was consolidating into fewer, larger concerns interested in managing a smaller number of bigger brands rather than a hugely diversified portfolio of more arcane products of specialist appeal, and so the drift towards standardisation was perhaps inevitable. Add to this the inflationary pressures of the 1970s, particularly the rise in the price of the heavy oil then widely used to fire stills, and the distillers were caught by falling demand and rising costs. Strongly branded, uniform blends seemed to offer a way out.

Then, however, as any fule kno, there was a reaction from the consumer. The early signs of this were the fortunes of smaller producers marketing their single malts with growing success (Glenfiddich, Glenmorangie and The Macallan, all three then still

independent, were early pioneers) and the formation of the Scotch
Malt Whisky Society, a sort of CAMRA for whisky, in 1983, though
it had existed for some years prior to that on an informal basis.

You didn't even, in fact, have to be a malt maniac to object to the
growing blandness of the whisky that was generally available, but it
probably helped. I don't think there is really very much question
now that the variety and range of taste of whisky produced in the
1970s and 1980s was greatly reduced from that seen in previous eras,
though, ironically, there is now a fashion for buying old bottles
from this period and extravagantly singing their praises.

Some of this was a good thing, as inconsistent and aberrant
whiskies were less frequently encountered, but, taken as a whole,
these are probably not whisky's finest or most memorable decades.
But even within the trade there were critics and heretics who
insisted on going their own way, regardless of wider trends, and
some commentators, most notably the late Michael Jackson, were
quick to point out the dangers in the path the industry had adopted.
Jackson was one of the earliest and most passionate advocates of
the pleasures of Islay whisky, so I believe that today he would be
pleased and not a little surprised by the range of whiskies which
it has proved possible to produce from plant that was designed to
create the type of uniformity he so roundly condemned.

In the event, the whisky industry has proved immensely
flexible and inventive in responding to changing consumer tastes
and the apparently inflexible production regime of the 1980s has
shown itself capable of producing a hitherto unimaginable amount
of varied whiskies, of high and consistent quality.

This is very well illustrated at Caol Ila, where both peated
and unpeated styles are produced to considerable acclaim. The
Johnnie Walker blenders are, one presumes, entirely happy, while
the single malt enthusiast is offered genuine choice and variety
from the output of this one distillery. And, as an unintentional

but happy consequence, it must be considerably more interesting and satisfying to work there when not on nature watch.

Both Bunnahabhain and Caol Ila are distilleries that have seen great changes: built to serve in an era of blended dominance, they have demonstrated a considerable ability to respond with diverse and interesting whiskies. Both, I think, suffer from their history and from their location at the currently less unfashionable end of the island. But recent fieldwork by the University of Reading has established that there were important Mesolithic and Palaeolithic settlements at Rubha Port an t-Seilich, just to the south of the present-day settlement. People have been settling here for thousands of years.

A flint-point arrowhead has been found there which is believed to be around 12,000 years old, making it the oldest known artefact from Islay and an object of European significance in understanding the late Ice Age hunter-gatherers who first colonised much of the Hebrides. It seems they lived on reindeer, red deer, roe deer, wild boar, badger, fish and hazelnuts—quite a healthy diet when you come to think about it. The excavations at Rubha Port and elsewhere on Islay are revealing more and more about the pre-history of the island. Along with the pioneering work being carried out on Orkney, we are being forced to rethink our view of the earliest settlers and the importance of the Hebrides and Orkney to Britain's history.

Port Askaig today is a relatively modest settlement. The Caol Ila distillery houses were favourably commented upon by Barnard, who claimed to envy the healthy lifestyle of the workers. I feel a degree of romantic, artistic licence was being employed, however, as I doubt he would have welcomed the hard physical labour involved in distilling back then.

Port Askaig was for many years the island's traditional point of arrival and departure, especially for drovers taking their beasts to market. They were ferried to Jura, walked then to Lagg and taken on another ferry to the Scottish mainland. Today, larger vessels arrive here from the mainland and the facilities at Port Askaig have been greatly improved in recent years.

The Scots baronial house that sits dramatically on the headland above the village is Dunlossit House, the long-term home of the Schroder banking family. Over the years the family have been generous supporters of the RNLI and the *Helmut Schroder of Dunlossit II* lifeboat may often be seen tied up by Port Askaig. More relevant perhaps to this book is the fact that the current owner of the house and estate, Baron Bruno von Schroder, was one of the early shareholders and supporters of the revived Bruichladdich Distillery. Said by Forbes magazine to be worth some $5 billion, he is also recorded as having donated £30,000 in March 2014 to the Better Together campaign in support of Scotland remaining part of the United Kingdom.

However, with those thoughts I fear I may be in danger of drifting into politics, which is never a terribly good idea. Let us turn our backs on Port Askaig, Palaeolithic settlers and all, and follow the lead of the good Baron Schroder to the western side of the island where may be found Bruichladdich and Kilchoman, which until one of the four proposed sites actually begins operations, is Islay's newest distillery.

STOP PRESS: Some welcome news has just arrived! Bunnahabhain have announced an £11 million improvement programme at the distillery. The road will be upgraded; the distillery and some warehouses are to be refurbished and new warehouses built and—best of all—the seafront cottages are to be restored and brought back into service for tourism and the distillery's guests. I wonder if they'll invite any writers....

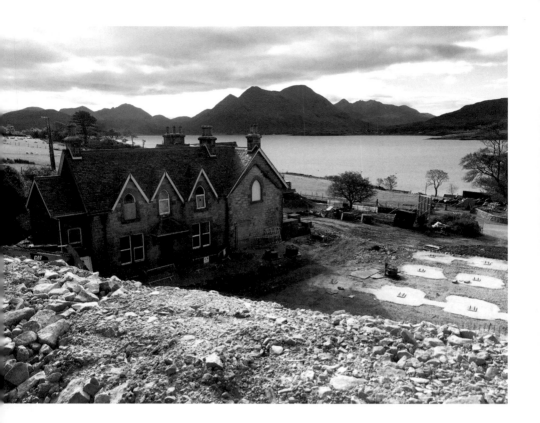

Top: The launch pads for Raasay Distillery due to open in late 2017; Skye in the background. *Bottom left:* The gravestone of Sam the faithful dog at the abandoned Bunavoneader Whaling Station on Harris. *Bottom right:* Smile! The enigmatically silent Scapa 'foghorn' on Orkney.

Top: Tobermory: the distillery staff wait to greet their visitors.
Bottom: 'Get it drunk!' Willie Cochrane, Jura Distillery.

op: The old drove jetty at Glas Uig, Islay where German U-boats would anchor for a spot of
heep rustling during both World Wars. *Bottom*: Ian Richardson's boatyard, Stromness; the
Orkney Historic Boatyard Society is a beneficiary of sponsorship from the Scapa Distillery.

Top: Skye: The distillery I thought would never be built: Torabhaig, September 2006.
Bottom: Oops! A decade on, Torabhaig under construction.

Whisky Hero: Ronnie Lee, weightlifter and the 'Man who Mends the Mills' in his workshop.

In memory of Alfred Barnard: a traffic jam on the A846, Islay.

Top: Islay: Visitors should arrive promptly at the Kildalton Cross 'café' if they require the legendary lemon drizzle cake. *Bottom:* Mark Reynier (right), then of Bruichladdich returns the Yellow Submarine to a somewhat chastened Ruler of the Queen's Navy.

Distillery workers' houses at Bunnahabhain with the Paps of Jura beyond. The village which was built together with the distillery, once housed its own school, reading room and accommodation for the Excise Officer.

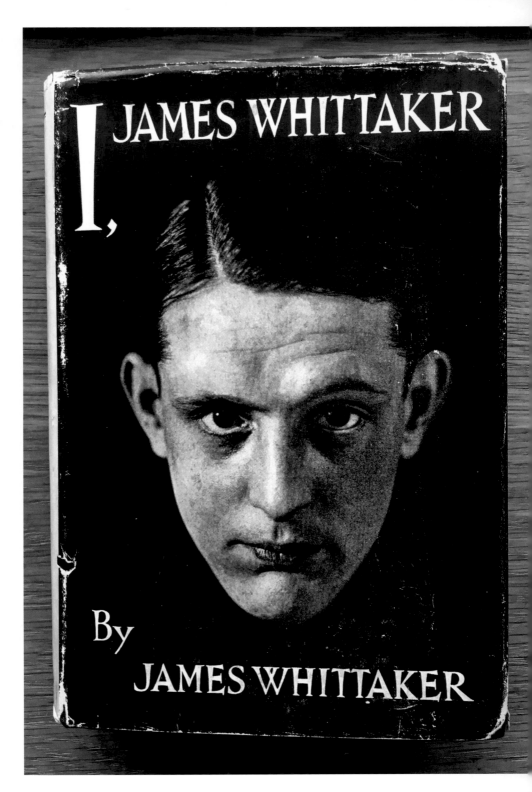

I, JAMES WHITTAKER

By

JAMES WHITTAKER

James Whittaker, a cooper's son who grew up in the village of Laphroaig, and whose memoir the distillery tried to ban.

Top: Lagavulin from the sea, with Dunyvaig Castle prominent. *Bottom*: Floor malting at Kilchoman on Islay, with an authentic grain handling and transport system in the foreground.

The Still House at Abhainn Dearg, Harris — the authentic island air of disarray, but a delightful time machine to a simpler age.

'There are flavours in it, insinuating and remote, from mountain torrents and the scanty soil on moorland rocks and slanting, rare sun-shafts' (Whisky, Aeneas MacDonald).

Islay: on an 'experiential' visit to Bowmore, a group of hipster cocktail barmen leave the confines of Shoreditch and learn how to cut peat. Tattoos were optional.

O. O. Whisky
OLD ORKNEY

Stromness Distillery, Orkney, Scotland.

Fortunately for the Old Orkney Brand Manager, social media had not yet been invented and this flagrant breach of all known advertising codes of practice passed unnoticed for around one hundred years.

6

Islay

The yellow submarine

AFTER LOOKING AT THEM FOR A WHILE, I made a decision about all those books about Islay. I could read them for research, and they do look lovely, but that seemed like quite hard work and in the end I put them all back on the shelf. They're there now, staring rather moodily at me and emitting a sort of smoky resentment. Perhaps, like a really peaty whisky, that will fade with age.

Bruichladdich is actually quite a good place for people like me (the unpeated freaks; can't see that catching on as a nickname) to start exploring Islay whisky. Not only was it historically quite a mild-mannered dram, the recent history of the distillery provides lots of amusement. Their irrepressible MD, the ebullient Mark Reynier, could be relied upon to provide a provocative quote to fill out an article or blog. He's moved on; Bruichladdich is a little quieter as a result, and drinks journalists have to work that much harder.

But those quotes were too good to let this pass without at least one example. For most of his tenure he enjoyed a distant and strained relationship with the rest of the whisky industry, writing a strongly opinionated blog in which he expressed himself pretty freely on all sorts of matters. I accused him once of 'grand-standing from a position of complete and total ignorance'—which he took in pretty good part (mine wasn't an entirely fair comment, but this is a man who enjoys a robust exchange of views)—and then suggested that he try to engage with the rest of the industry.

You have to admire the candour of his reply: 'I have absolutely no desire to "engage" the rest of the industry, a pointless exercise as there is little I can learn from them and vice versa. Besides, I am perfectly happy "sniping from the side-lines" thank you very much.'

He once memorably summed up the Scotch whisky industry's marketing efforts as demonstrating 'a staggering lack of vision, insecurity, timidity, sparsity of imagination, fear of failure, [and] paucity of thought and integrity', from which you may take it that he wasn't overly impressed. Not much fear there, but quite a lot of loathing. It's my impression that the feeling was mutual.

But actually, I know quite a good story about Mark and Bruichladdich.

Founded in 1881 (something of an annus mirabilis on Islay; as we've already noticed, Bunnahabhain was also built then), it passed through a bewildering variety of hands before being mothballed in 1995, not for the first time. On 19th December 2000, however, it was bought by a small independent company financed by bank loans and a group of twenty-one private individuals, some from Islay and some quite new to the place.

The team was headed by Mark and his colleagues Gordon Wright and Simon Coughlin. In their previous lives, they had been wine merchants and independent whisky bottlers. They've gone on about the re-opening ever since. But here's the story, as Mark Reynier told it to me.

Some years prior to buying Bruichladdich and with no such idea then in his mind (or so he maintains), Reynier made his first visit to Islay and drove past the distillery, the stout iron gates of which were firmly padlocked. No whisky was being made; everything was shuttered and bolted. But being possessed of a restless and inquisitive turn of mind, he stopped and looked into the courtyard where he happened to see a worker, presumably some sort of watchman.

He beckoned him over. 'I was wondering if I could look around the distillery,' he enquired.

'No!', came the reply. 'You can fuck off.'

Shortly after his team acquired the business there were apparently a number of personnel changes. And that story, funny though it is, captures something of the anarchic Bruichladdich spirit. Right from December 2000 they cast themselves as the enfants terribles of the whisky industry: you were either with them, or against them. There was no fence to sit on.

Now I'm not actually sure that very many members of the industry cared all that much or, at first, even really noticed. But with the Bruichladdich team's undoubted genius at marketing, a growing number of malt enthusiast consumers did, and the maverick approach greatly appealed to a certain rebellious streak in a decent number of drinkers.

At first, getting the distillery going was very hard work. Because the planned date for the purchase slipped back and back, the vital Christmas sales season was missed and it was necessary to return, cap in hand, to the shareholders to request more money. And then a second cash call was necessary; yet the remarkable thing is that all the shareholders supported the refinancing.

It was May 2001 before the distillery was working again. The initial releases were more than a little confusing unless you followed Bruichladdich very closely. To some extent this was driven by their obsessive insistence on doing things differently from the rest of the whisky industry; while it's also true to say that the patchy quality of the stocks they inherited with the distillery pushed them towards a bewildering range of expressions and finishes.

Not that you were supposed to refer to 'finishes'. Bruichladdich insisted that this was ACE-ing (Additional Cask Evolution). The term proved controversial and was certainly not the revolutionary and radically different method of enhancing whisky that some of the more excitable commentary might have implied. Noticeably, as the old stock has been depleted and post-2001 distillation whisky has emerged, the use of the ACE-ing technique has fallen away.

Reynier always made a great play of the fact that Bruichladdich didn't do marketing. That, of course, is nonsense. Their marketing consisted, as often as not, of inspired PR based around irreverent products—this was marketing masquerading as 'anti-marketing' in order to get talked about, and it worked very well.

Often they were deliberately provocative, stirring up the established order to draw attention to their renegade pose and self-proclaimed outsider status.

Today they might prefer the term 'progressive'—in fact they've painted this legend on the wall facing the distillery car park just so you don't forget.

But when they weren't taking themselves too seriously, Bruichladdich had some of their finest moments.

An example is the X4 series, which was inspired by the writings of an old chum Martin Martin (we will meet him from time to time) who in 1703 described a quadruple-distilled spirit which he called Usquebaugh-baul. As he relates, this 'at first taste affects all the Members of the Body; two spoonfuls of this last Liquor is a sufficient Dose; and if any Man exceeds this, it would presently stop his Breath, and endanger his life.' Interestingly, he suggests it was made with oats.

Curious, thought Bruichladdich, translating Usquebaugh-baul as 'Perilous Whisky' and wondering what it would taste like. So, they made some, and it sold very well and attracted lots of commentary—lots—including a po-faced condemnation by the Scotch Whisky Association who thought it 'irresponsible'. All of this, of course, was simply what Bruichladdich had expected and secretly wanted, as it merely increased the notoriety of the product and drew people to it.

At its full strength of around 90% I daresay that more than a few spoonfuls would cause a sharp intake of breath at the very least, but it was actually bottled and sold at the more reasonable 63.5%. Still eye-watering enough, but no different from any number of apparently acceptable cask-strength releases. It must have been entirely coincidental that Bruichladdich was not then a member of the SWA and was a voluble critic of it.

Mind you, at the full 90% strength it was possible to use it to power a sports car to more than 100mph, as motoring journalist James May proved on the BBC, going on to note however, that at £120 a gallon it was better to drink it. May achieved a 'moment of hope and great purity' on trying it and, in possibly the greatest tasting note ever written, Charles MacLean memorably described the nose as 'not cooking oil. Not diesel oil. Sewing machine oil.' Not that Bruichladdich cared, for if you don't do marketing, all publicity is good publicity.

X4 was one of their more extreme WMDs, Whisky of Mass Dis-tinction, or, as sometimes styled, Weapon of Mass Drunkenness. Their first commemorated Orwell's 1984, but perhaps the most famous was WMD II—The Yellow Submarine.

Bruichladdich had already enjoyed a massive boost when it was revealed, with perhaps a little embellishment, that the

US Defence Threat Reduction Agency had been monitoring the distillery's webcams. Alerted by the movement of old equipment from the Inverleven distillery on the Clyde, which was being towed on a barge to Bruichladdich, the DTRA somehow deduced that the equipment could be used to produce 'Weapons of Mass Destruction', then something of an obsession in the media.

As the media clamour died down, Bruichladdich moved quickly to obtain further advantage with a special bottling. Here their size is an advantage: where a larger, more process-bound business would take weeks or months debating the possibility and considering the risks to their corporate reputation, the Bruichladdich team were able to get up one morning and decide to do it. So they at once set to designing a label and launching the product while the opportunity was still current.

A further occasion to exploit the coverage then presented itself as, shortly after the first WMD fun, a bright yellow Ministry of Defence unmanned mini-submarine was lost from a minesweeper off Islay. Salvaged by a local fisherman, the Royal Navy at first denied it belonged to them. The Bruichladdich publicity machine leapt into action, a further bottling was rapidly developed, the world's media grabbed the story with grateful hands and, by the time HMS Blyth slipped into Port Ellen to reclaim their submersible, the distillery were ready with a case of whisky to present to the grateful, if discomfited, mariners. The bottling rapidly became a collectable; retailed at launch for £45, you might easily pay ten times that should a bottle become available today.

And that, you might think, was that. Except that the yellow submarine became the gift that keeps on giving. Quite extraordinarily, by 2016 the Royal Navy had decided that it was obsolete so it was offered for sale. Hard though I find it to credit this, they put it on eBay.

The distillery bought it, of course, and at the 2016 Islay Festival it was installed to wild acclaim as a display piece in the Bruichladdich courtyard. Two bottles of the original release were auctioned, raising some £3,000 for the Islay & Jura Sick Children's Fund. And everyone present got a dram.

Sadly, we hear a little less of such wonderful nonsense from Bruichladdich these days, and they seem to have quietly dropped their Gaelic motto 'Clachan a Choin' (it translates as 'the dog's bollocks'). The company has been sold, the ever-patient shareholders finally deciding that a £58 million offer from Rémy Cointreau was sufficient to relinquish the company's much-vaunted independence; Mark Reynier has moved on to a new project in Ireland; and Simon Coughlin, who took over as MD, has been promoted to a bigger job managing all of the growing Rémy whisky portfolio.

But much remains. Despite their new corporate ownership, Bruichladdich remain outside the SWA and they maintain a staunchly independent point of view, even if these days it is expressed with greater diplomacy. Less effort is expended on the deliberately provocative, but their stance remains both thought-provoking and intellectually stimulating.

It's timely at this point to observe that several of their original cardinal principles, which seemed so very radical just fifteen years ago, have now been more widely adopted across the industry. Though hardly universal, their insistence on bottling single malt whisky without chill filtration or added colour has been taken up by many of their competitors, and for all that ACE-ing may have been a necessary step to enhance less than exciting whisky from the previous regime, it cannot be denied that Bruichladdich's work here resonated with enthusiast consumers and has influenced the rest of the industry's view.

Some of their more outlandish claims have been dropped. For some time, the distillery's website was pleased to describe their equipment as 'pre-industrial', an absurd proposition for machinery installed in 1881. I think founder William Harvey would be quite offended: he built a state-of-the-art facility for the period, well into the Second Industrial Revolution. Barnard lavished praise on the 'powerful steam engine' he saw there. I don't think this is a trivial point; Bruichladdich make some considerable claims for themselves and deserve to be taken seriously.

They also proclaimed loudly that they would be rebuilding and re-opening the old distillery in Port Charlotte (Lochindaal), which was closed in 1929 and largely demolished. It enjoyed a one-hundred-year run, under a considerable number of owners, which by the standards of many a small distillery wasn't such a short life.

Bruichladdich's more heavily peated style (apart from the Octomore expression) is marketed as Port Charlotte as a tribute to the lost distillery. The plan to recreate the distillery, always an ambitious goal, was announced by Mark Reynier in early 2007, catching his colleagues completely unaware, as on occasion happened.

Back in those happy days when Bruichladdich needed money desperately and was prepared to sell casks of new-make spirit to private buyers, I bought a cask of Port Charlotte. It was my way of showing tangible support for the project, which I thought exciting and innovative. It also appealed to me to have a cask of whisky with my name on it.

It was distilled on 12th June 2002 and, needing cash as one so often does, bottled on 5th November 2007 and sold via the Royal Mile Whiskies shop. I seem to recall that a 50cl bottle at 46% abv retailed at £39, so there was a sharp intake of breath at Buxton Towers when I discovered one on sale recently on a well-known retailer's website for £225!

Since the 2007 announcement of the old distillery's improbable renaissance little or nothing has happened, nor will it.

Bruichladdich was not then operating at full capacity, so other than a romantic attachment to the idea of another distillery, there was no real argument for building a further plant. The new owners have spent freely on expanding output at Bruichladdich, constructing new warehouses, improving the offices and funding a sustained global sales and marketing campaign: it is highly unlikely that they would take kindly to any proposal to rebuild Port Charlotte. Apart from anything else they have been fully engaged in buying small distilleries in France and the USA. The payback time on the investment involved in a new Islay distillery would be unrealistic, so I fear the dream must be considered as no more than that.

It was fun to be a cask owner, albeit only briefly. At that time it was possible to go to the distillery to visit your cask and sample it for quality assurance purposes. Sadly, it has been determined some while since that the manpower involved in this (together with the onerous paperwork required by HMRC) makes it no longer viable. And, of course, they no longer need your money, and require all the whisky themselves for future demand. That's reasonable enough, I suppose, and it was fun while it lasted, but I feel something's been lost. To be fair to Bruichladdich, virtually all these offers of private cask sales have been withdrawn and those that remain seem exorbitantly expensive to me.

But it was fun — I envied the Dutch guys who owned the cask I saw on one of my visits. Walking through one of the very evocative traditional dunnage warehouses I spotted a cask that seemed a little out of the ordinary. On closer inspection it was festooned with streamers; several birthday cards were pinned to the cask end and there was a party hat perched on top of the cask.

'What's going on here?' I asked my guide.

'Oh, that one belongs to some Dutch guys. Once a year they come over on the anniversary of the filling date, sing "Happy

Birthday" to the barrel, taste the whisky and then go off to play golf. And have a few drams, I expect.'

Now I've never seen anything like that anywhere else, before or since, and I think that's a great story. What loyalty. What identification between consumer and product. Those happy Dutch guys will be the best ambassadors any distillery ever had, and they paid for the privilege.

And how smart of Bruichladdich to allow them to let off a few party poppers in the warehouse. Almost everywhere else, some health and safety jobsworth would have stepped in to stop the party, citing some perceived 'explosion risk'. From a party popper.

I don't know if the cask is still there, but I hope so. It's as vivid a demonstration of what made Bruichladdich different as any of their madcap expressions, passionate press releases or diatribes against the giants of the whisky industry. There, in a few party streamers and some birthday cards, you begin to understand why people love this distillery.

Today they've matured, as well as aged; they've calmed down the rhetoric and, you feel, are no longer looking to offend just for the sake of being the bad boys, the bohemian iconoclasts of whisky. That role, incidentally, seems about to be taken up by Lone Wolf, the distilling arm of BrewDog, so stand by for some puerile and tedious posturing from the direction of Ellon in Aberdeenshire. BrewDog, incidentally, are so 'punk' that their founders happily accepted MBEs and have sold a 23% share in the company to a San Francisco venture capital fund, valuing their business at a staggering £1bn. Not at all bad for something founded in a garage in 2007 but not, one would have imagined, particularly 'punk'.

Bruichladdich are still throwing out some serious challenges to the established order though. In a world in which malting barley is sourced from all round the globe largely on the basis of the anticipated yield of spirit per ton of grain set against the price of the harvest, Bruichladdich are increasingly obsessed with provenance and terroir. As they express it, 'We believe that Islay whisky should have an authenticity derived from where it is distilled and where it is matured... From the philosophies of those who distil it. A sense of place, of terroir that speaks of the land, the barley and water from which it was made, and of the human soul that gave it life.'

The links to the wine trade background of their founders are very clear. They make the point, firmly, that Scotch whisky is called Scotch for a reason and, for them, that requires nothing less than the use of 100% Scottish barley in everything they produce. It may come as a surprise that not all distillers are equally demanding. Barley is sourced internationally, though naturally this isn't something that is widely discussed.

Challenged on this, other distillers will maintain that the origin and, indeed, variety of the barley they use makes little or no difference to the final product. The cask selection, contribution of the wood and skill of the blender is more important, it's said, to the eventual taste. Any variation in flavour due to the barley is swamped by these later, more forceful influences. And, yes, there is something in that, especially if your product is a high-volume, mass-market blended whisky designed for mixing or serving over ice. Or if you sell into the cheaper end of the market, where every penny of cost in raw materials must be carefully monitored and fully justified.

But, fortunately, this is not Bruichladdich's world and they can ignore such concerns. In this, they are not alone. Many distillers produce premium and super-premium products, especially single malt whiskies, where a few additional pennies per litre in the cost of the spirit will have little or no influence on the final shelf price. However, though others may enjoy a similar premium positioning, few, if any, are as single-minded on this subject. Terroir is not a concept that is frequently expressed in the world of Scotch whisky; indeed Bruichladdich would maintain that it is 'much maligned by those for whom it is commercially inconvenient'. That may be so, though I suspect that they may protest just a little too loudly; be a trifle over-sensitive in making their case.

For them, however, micro-provenance is all. They label and trace, parcel by parcel, different barley varieties, farms, even the different fields that go into their maturing whisky with equal care, attention and obsession going into cask selection and maturation. This has now reached the point where they are releasing whiskies traceable back to individual fields and harvests.

An extreme example of this is their annual release of a Bere Barley edition, using grain specially grown for them on Orkney under the supervision of the Agronomy Institute. Bere is low-yielding, and hard to work with in the distillery. Indeed, when Bruichladdich first attempted to distil using bere it presented huge physical problems, clogging together and causing the Victorian cast-iron rake and plough mash tun to seize into a solid mess of hot porridge. But they believe it worth it for the final product, convinced as they are that barley variety and origin make a critical difference to whisky's flavour.

Of course, scale helps. A larger brand from a larger distillery could not do this, indeed would not want to, and larger producers loudly maintain that the barley variety makes no difference to the taste. It's a question of a radically different ethos and belief system that makes Bruichladdich stand apart.

Not everyone gets it and it requires some thought, effort and intellectual input from the drinker. Not every drinker wants to do that and nor, perhaps, should it be asked of them.

Sometimes it's nice just to sit down with a whisky and not have to think about it—that's probably not a Bruichladdich moment though. But when you do want to think about it, then Bruichladdich gives you plenty on which to ponder.

The remarkable thing about the Rémy Cointreau takeover is how few things have changed, though one can see the obvious cultural familiarity of the new French owners with the Bruichladdich philosophies and approach. Most of the original team remain in place, or have retired naturally, as they would have done in any event, handing over to their planned successors. Some money has been pumped into the local economy, as most of the original employees were shareholders, and Rémy have spent on the buildings and new warehouses. But Bruichladdich remains very much the operation that it has been for the past fifteen years—quieter perhaps, but more confident and assured of its purpose and place in the world of whisky.

Strangely, I put a lot of that down to their gin. Much of the equipment that excited the interest of the US spooks left the site along with the then MD Mark Reynier. The old Inverleven stills that stood proudly outside Bruichladdich (complete with the welly boots poking out of the neck of the still) have gone. They were destined for the putative Port Charlotte distillery but have, instead, found a new life in Waterford, Ireland where Reynier is making Irish whiskey.

But one important piece of equipment remains and is busy earning its keep. This is 'Ugly Betty', a rare Lomond still that has been installed in the cramped Bruichladdich still room where it makes The Botanist, the one and so far only Islay gin.

The Lomond still was a relatively short-lived innovation. Invented in 1955 by engineers at the Hiram Walker company, it was an attempt to produce a flexible pot still that was versatile enough to produce very different styles of spirit from the same apparatus. This was to be achieved by means of perforated plates fitted into the neck of the still. The position and angle of these could be altered, or one or more plates could be removed, thus increasing or decreasing the amount of reflux (spirit which condensed and ran back into the still before passing over the lyne arm to the condensers) and accordingly influencing the final spirit character.

The theory is sound. Unfortunately, in practice, the Lomond design did not work as well as predicted and the influence on spirit character was less than had been hoped for. Few were built as it was recognised as an evolutionary failure. Fewer still have survived. The Loch Lomond distillery at Alexandria have several that are still working, though, and one remaining Lomond still serves as a wash still at Scapa, on Orkney, though with plates removed I maintain that it's closer to a conventional wash still with a cylindrical neck than a fully operational Lomond still. Curious that it should end up on an island though.

The very first example, however, was installed at the Inverleven distillery in 1959 — the same derelict plant that the Bruichladdich team dismantled and shipped to Islay on those troublesome barges — so eventually it found a home where it was loved.

And there it found a new life as a gin still. It's not, I might add, the prettiest or most elegant still you'll ever see. Styling it 'Ugly Betty' seems a trifle harsh though, for out of this remnant of whisky's history, forgotten and forlorn in a silent and unheralded distillery, fated to end in ignominy as so much copper scrap, there runs today a spirit of rare delicacy and refinement that has enjoyed surprising success out of all proportion to the initial expectations.

Gin is not popularly associated with Scotland so it comes as something of a surprise to see it made here. But, in actual fact, Scotland is a major gin producer. Apart from the small craft operations that are springing up across the country, several major brands are distilled here — Gordon's and Tanqueray in Fife and Hendrick's in Girvan probably make up close to two thirds of all the gin distilled in the UK. The only major brands left in England are Beefeater and Bombay Sapphire, with the Greenall distillery in

Warrington producing that label and some supermarket products.

And here's another surprise: so successful has The Botanist been that Bruichladdich will sell, by volume, more gin this year than whisky. That's a remarkable statistic that speaks volumes for the worldwide renaissance of gin sales and the quality of this product. It's also a sobering statistic for all the ardent single malt aficionados who adore Bruichladdich's whisky—what if gin eventually takes over the distillery?

Well, we're very far from that. Whisky is too deeply rooted here for gin to take its place but, for all that, what has been achieved with The Botanist is quite exceptional, even at a time when gin is enjoying an unprecedented boom.

Don't be fooled into thinking of this as any less a product. Great gin is not easy to make and, unless I'm mistaken, The Botanist is a great gin. It's styled Islay Gin, naturally, in acknowledgement of the twenty-two local botanicals that contribute to its distinctive flavour. Here we return to Bruichladdich's obsession with terroir and local identity, as sourcing these ingredients on the island is as essential to this product as Scottish barley is to their whisky.

The success of The Botanist may have come as a pleasant surprise to Rémy Cointreau—indeed, well-informed sources have suggested to me that they would have been perfectly happy to see the brand leave with the former MD—but it would surely be impossible to prise it from their hands now. Not only is it a highly profitable global success, but it spreads the Rémy portfolio into a category where they were previously unrepresented.

Adapting the Lomond still to the production of gin seems to me to personify the 'can do' attitude that characterised Bruichladdich for so long. An attitude cultivated of slight paranoia that an uncaring world was set against them, that carried the initial team through the hard times of the distillery's loss-making first few years. It's worth remembering that hard decisions were taken in those early days: there were cutbacks and some of the founding team departed to cut costs. It was not always clear that Bruichladdich was going to be the success that it has become, and despite having had the very good fortune to have been in the right place at the right time with both single malt whisky and more recently gin, theirs is not an achievement to be taken lightly.

Much the same could be said of their near neighbour Kilchoman. To reach this little farmhouse distillery, currently the newest and smallest on the island, it's necessary to cross over the low hills behind Bruichladdich to the other side of Islay. You could walk, I suppose, but by far the most convenient way to get there is by taking the car. So, leave Bruichladdich by the A847 as if to return to Bridgend. Following the coastline, just before the road turns to the right there is a sharp left-hand turn to be made onto the B8018. Take care: there are attractive views all round but on this single-track route there is always someone else coming the other way! This is a surprisingly hazardous six miles and the unwary traveller used to city traffic can easily be caught unawares by a tractor, sheep trailer or giant camper van full of bearded Germans.

In fact, Islay tends to be overrun by bearded whisky enthusiasts these days, often from Germany, Holland or the Nordic countries. It was in those territories that the Islay style really started to pick up momentum and it seems that our Continental friends, especially from Northern Europe, can't get enough of the stuff—the peatier the better. During the week of madness better known as the Fèis Ìle, Islay is home to more mobile homes and camper vans per square mile that anywhere else on the surface of the planet. Good old CalMac even lay on additional ferries, so great is the demand.

But assuming you've managed to avoid the traffic, the B8018 will take you to a signpost for Kilchoman distillery, which was established in old buildings at Rockside farm as recently as 2005. That's more than a decade ago now, so I find myself mildly reeling in astonishment that no one's written a book about it yet. I'm sure there are plenty of stories.

There were certainly plenty of sceptics when the plan was first announced. Hard though it is to recall now as we gambol in the new-found prosperity of our twenty-first century whisky boom, building the first new distillery on Islay in 125 years was not an immediately obvious proposition, nor a sure-fire success.

In fact, Kilchoman wasn't just the first distillery to be built on Islay in a very long time, but a pioneer for an international wave of artisanal distilleries. Between 1980 and 2005 there had been precious few new distilleries opened in Scotland, while more than twenty had closed.

The newbies presented a curious bunch: Kininvie, a low-profile Speyside distillery in the William Grant & Son stable; Arran, of which I have written here; Glengyle, a modest Campbeltown revival; and Kilchoman. Around the same time as Kilchoman was under construction, other plans were floated such as Ladybank and Barra, neither of which have yet commenced operations (some building work started at Ladybank, but was never completed, apparently due to the company being severely underfunded).

However, while Kilchoman opened in a blaze of publicity and has continued to maintain a high profile ever since, it presents a curious contrast to another small farm distillery opened in December 2005. That was Daftmill in Fife, and while it continues in operation, no whisky has yet been released, as the owners, Francis and Ian Cuthbert, combine distilling with their farm operations and appear content to sit on the whisky that has been made until such time as they deem it ready.

Such a luxury was never available to the founder of Kilchoman, Anthony Wills. Delivering his vision, then seen by many as cranky and idiosyncratic, of a small farmhouse distillery incorporating the grass-roots traditions of Scotch whisky has proved an onerous and demanding task, not to mention an expensive one, fraught with difficulties. I come back to my advice about opening a distillery. If the idea strikes, go and lie down until the mood passes: it will take at least twice as long and cost at least three times what you think and any number of ostensibly 'helpful' public agencies will put unimaginable obstacles in your way, while all the time insisting on their sincere support for your project. Further meetings will, however, be required at which further studies will be commissioned, leading to further meetings to review the findings.

As no one has yet thought fit to write a book about Kilchoman—scandalous omission—a few more detailed notes may be helpful. The distillery is remote, even by the standards of Islay, and definitely requires a special trip. Having said that, it's only a short distance from a long sandy beach so all is not lost if the family object.

The setting perfectly illustrates the integration of farm and distillery, with all the barley used grown on Islay (most of it on the distillery's own farm), malted in the distillery's own maltings on Islay, matured on Islay and bottled on Islay. The plant is all quite modest in scale but, with a few process improvements and the input of a skilled and experienced distiller, it has proved possible to expand annual production from an initial 90,000 litres annually to over 200,000 litres and it is thought possible that up to 250,000 could eventually be produced.

Originally, Kilchoman was to be operated in parallel with the adjacent farm which at the time raised beef cattle and ran a small riding school. However, there were disagreements with the farm owner, who held the lease for the distillery and for a while both were run entirely separately and not entirely harmoniously—a stressful situation for both parties. This uncomfortable co-existence would eventually have threatened Kilchoman's long-term future and certainly constrained investment and development on the site. In June 2015, however, the farm owner decided to put the farm on the market and it proved possible for Kilchoman to acquire the farmhouse, land and buildings, thus taking control of the entire complex.

Today, the visitor will find a small but welcoming shop, a friendly café and the distillery, which is open for tours and tastings. Walking round the site with Anthony Wills — a privilege of authorship, by the way, for Kilchoman is not so small that everyone gets a personal tour from the managing director — I was struck by the contrast between the site bustling with visitors and Kilchoman as I first saw it.

Then, around the turn of the millennium, all there was to see was an ambitious, but slightly faded, sign on the wall of a semi-derelict farm steading suggesting that this was to be the still house. Frankly I didn't believe it; the site was so remote, the condition of the buildings so parlous and the whole place so sleepy that it seemed inconceivable that anyone seriously proposed a distillery here.

Five years later I was happy to be proved wrong, but the stress, cost and worry involved in delivering Kilchoman showed vividly on Anthony Wills at the opening, when he was close to being overcome with emotion. Distilleries do this to people and I'm not always certain that the price is worth paying; while today it seems a great deal easier (not 'easy', note, simply 'easier') to raise the finance for a new distillery, that's because we're in a remarkable era which combines low interest rates, global capital chasing the promise of decent returns and the current fashionability of whisky. Just like the 1890s in fact, and that whisky boom ended very badly indeed.

However, having been proved wrong before, let's hope that the sign at Gartbreck on the old farm buildings there will be as favourable an omen for success and prosperity as the one I scornfully dismissed all those years ago at Kilchoman.

A good proportion of the visitors to Kilchoman are, of course, aspiring distillers in their own right, with their own hopes, plans and dreams for their own distillery. Some make a surreptitious visit, though their questions often give them away, while others are more honest and open, assuming that advice will be freely given. From the numbers involved it's evident that there are still a remarkable number of optimists left in this world. I only hope that not too many dreams are shattered at too great a cost. As any brief study of whisky's history will reveal, the past is littered with failed distillery projects, not least on Islay.

7

Islay

Unfit for human consumption

'JUST POUR IT AWAY AND GIVE THEM THEIR MONEY BACK,'
said the portly gentleman firmly. 'Nothing that tastes like this was
meant for human consumption.'

The year was 1977 and I was standing in the Wines & Spirits
cellar of the Devenish brewery in Weymouth. The speaker was
Bob Carter, then Wines & Spirits Director of that business. An
experienced judge of wines and spirits, and well respected in
the trade for his discernment and knowledge, he was describing
a bottle of Lagavulin which had been returned by one of the
brewery's tied houses as 'off'.

What was I doing there? Well, despite being a very junior
member of the brewery's marketing team, I was the only Scot in
the building and so had been called to give the benefit of my local

knowledge and help adjudicate on the apparently errant bottle. The general assumption was that it had been filled in error with cleaning fluid, or some such noxious liquid, but that a Scotsman would know for sure.

I had only recently left university and, whatever students drink today, undergraduates in the 1970s did not generally consume single malt whisky. However, I did have some idea what Lagavulin was supposed to taste like so, on being asked my opinion, ventured tentatively that it was okay. You have to appreciate that I was quite the subordinate and Mr Carter, as he was universally referred to, a particularly forceful character.

He was not impressed by my impertinence and promptly issued his magisterial judgement, after which I was sent back to my office in some disgrace. I do not recall my opinion being sought again. Whether the bottle was flushed down the sink or returned to the distillers I never knew. I certainly didn't enquire. Incidentally today that bottle would be worth around £1,500.

Two things stand out about that story. Firstly, that forty years ago, a serious and experienced drinks trade executive would have no idea what Lagavulin should taste like and, secondly, that that bottle—it was then a twelve-year-old expression—carried the legend 'established 1742'. I remember it very well, because the Devenish Weymouth brewery was also founded in that same year and one of my very first jobs there was to write and produce a pamphlet on the company's history, so the coincidence of the dates struck me very strongly.

The suggestion that Lagavulin was established in 1742 and that the proprietors, White Horse Distillers, were confident enough of the dates to emblazon it prominently on their packaging in the 1970s is interesting as, during 2016, the current owners lavishly celebrated the distillery's 200th anniversary. I do not require a calculator to determine that they thus believe 1816 to be the year it was founded and that is curious, for why would anyone—other than a Hollywood actress of the old school, perhaps—wish to forget seventy-four years of their history? After all, heritage and provenance are powerful tools in the marketing of malt whisky and they could surely have waited a year to celebrate a 275th anniversary in 2017.

The 1742 date appears in Alfred Barnard's account of the distillery, but even he qualifies this as only being true to 'a certain extent'. He goes on to note that at that period there were 'ten small and separate smuggling bothies for the manufacture of "moonlight"' (i.e. contraband whisky). These clandestine operations may, in fact, be dated to 1631 or even earlier, but 1816 has the merit of being firmly and definitively the date when the first legal distillery was established by one John Johnston, whose family had other distilling interests on Islay.

It's heartening, I think, to see a little more regard for historical accuracy in whisky's marketing, albeit at the cost of some romance.

Lagavulin today is a powerhouse. Since 1989 there has been a policy to bottle it at sixteen years of age as opposed to the twelve years that prevailed previously. It's an improvement, I feel, and it seems that the locals agree: if you study the brand's promotional literature they are alleged to hold that 'time takes out the fire, but leaves in the warmth'. If it's in the promotional literature it must be true. I'm sure they say it all the time.

However, fire vs warmth was demonstrated, perhaps not wholly intentionally, during the 2016 anniversary celebrations when the most widely available commemorative bottling was of an eight-year-old expression, inspired by the experience of our old travelling companion Alfred Barnard. He apparently tried such an aged whisky at the distillery during a visit there in the 1880s and pronounced it 'exceptionally fine'. Mind you, it would have been unusually old for that time and so may have presented a favourable contrast with the younger spirit that was typically drunk before the requirement to age whisky was passed into law.

Bottled at 48% abv it is certainly powerful stuff. I tend to share the view of a very well-known whisky writer who, at a Diageo event to conclude the anniversary year, described it as 'coarse'. As it was a private event, he or she must remain anonymous, but I think it a just and fair appraisal. What Mr Carter would make of it I can hardly imagine.

At that same event two other very special commemorative drams were served—the twenty-five-year-old release and the one-off single cask bottling that concluded the year. There were just 8,000 bottles of the former whisky available, but even at £800 they sold out remarkably quickly. Having had the good fortune to taste it I will confess to growing increasingly impressed with this spirit as it grows older, helped in this case by its long maturation in sherry casks.

However, excellent though it was, it was over-shadowed by what is surely the ultimate Lagavulin—one which, sadly, will spend much of its life as a traded commodity, appearing on whisky auction sites at ever more inflated prices. This is the 1991 single cask charity bottling (cask strength at 52.7% abv), of which just 520 bottles were available.

Remarkably, they were offered for sale by ballot, the 'prize' in the ballot being the right to purchase a bottle for £1,494 which, by current standards, seems almost cheap. The ballot was, of course, massively over-subscribed—it turns out that more than 7,000 people wanted a bottle—presumably hoping to immediately flip it at auction to a wealthy collector or 'investor'.

In its way, this is both magnificent and poignantly tragic. It's magnificent because the sale was so organised that the entire proceeds received by Diageo (£580,000) were donated to Islay charities, but tragic in that I suspect very little of this superb whisky will ever be drunk. Most will end up in collections or, worse still, 'portfolios'; there to be gloated over by the Gollum-like tribe of whisky 'investors' who have so blighted the whisky market in recent years.

These Philistines talk of whisky indexes, of yields, of rates of return as if whisky were nothing more than a basket of commodities—pig belly futures, let us imagine—traded on some exchange in the expectation of some petty gain. They protest that they 'love whisky', but I fear they love money more and have lost sight of the fact that by confining their prizes to a locked display cabinet they have spurned the labour of all those who worked to create the whisky and whose efforts are meaningless until the liquid is drunk.

Whisky only attains meaning and significance, and only reaches its apotheosis by the very process of its destruction by the drinker. Until then, the crystal bottles and polished wood boxes of these trophies might as well contain cold tea. 'Tak aff your dram', as the poet would have it, and tell me nothing of your spreadsheets.

Please excuse the rant. The good news is that various worthy causes on Islay will share well over half a million pounds and even in our present inflationary age that's a significant sum on a small island. Most of the money, some £310,000, is going to a body called Islay Heritage which will support archaeological work, education and interpretation for local people and visitors alike. Compared to, say, Orkney, Islay's history is not particularly well understood and, as work such as the dig at Rubha Port has demonstrated, there is much of great interest and value that remains to be discovered, analysed and preserved.

Other grants have gone to the MacTaggart Leisure Centre and Cyber Café; the Finlaggan Trust; the RSPB; Islay Arts; and the Islay Festival. One might, if feeling just a little jaded and cynical, detect a degree of self-interest here: Islay Arts promote the Islay Jazz

Festival, which has Lagavulin as the title sponsor, and a significant part of Islay Heritage's grant will be spent on work at Dunyvaig Castle which is, coincidentally enough, conveniently adjacent to Lagavulin distillery. Charity begins at home, one might conclude, but even if that's so, these are both important causes and we should remember that as a commercial business with shareholders Diageo didn't have to do any of this. So I commend them for it, with the hope that at least some of the lucky purchasers will drink their bottle. I can, with confidence, say it is exceptionally fine.

Incidentally, if like me you are of an enquiring turn of mind you may already have calculated that the total revenue from 520 bottles sold at £1,494 each comes to £776,800. It's a lot of money, but considerably more that the charitable donation of £580,000. So what has happened to the 'missing' £196,880?

I knew you'd want to know, so I enquired on your behalf and curiously, at the time of writing, it seems I am the first and only person to ask. What this tells us about the legion of writers, journalists and bloggers who faithfully reprinted the press release without checking the arithmetic I hesitate to suggest. However, the answer is that VAT alone accounts for nearly £130,000 and, of course, the government also takes a duty charge. Then there are credit card commissions and the cost of the website design, general administration and managing the ballot and so on and so forth. Before you know it a cool 25% of the total sales value has disappeared. Makes you think. I suppose someone has to pay for our Trident missiles, HS2 and that tunnel under Stonehenge.

In actual fact, not all the bottles went into the ballot for sale. Two were destined for corporate archive collections and the first numbered bottle was held back to be auctioned to raise yet more money. Then I was greatly flattered to be asked as one of three whisky writers to contribute tasting notes to a leather-bound booklet that was to accompany the ultimate bottle. I was generously offered a fee but thought it churlish to ask for anything more than a sample. Hoped it might have been larger, but, as they insisted on handwritten notes, I got to use another pen. (The excitement just never stops...)

For this critical task, I deployed a second Visconti, one from their Opera range: a dashed good-looking fellow, something of a gigolo perhaps, with a blue resin body stippled with splashes and dots of red and light blue, which glint metallically under my desk light. The body is faceted in octagonal form but with the facets unequal in size. It fits very comfortably in the hand but, regrettably, reminds me slightly of the Italian sports cars of my youth; looking superb but performing only fitfully. Annoyingly, the ink does not flow evenly onto the nib and from time to time it's necessary to open the body of the pen and by screwing down the plunger on the ink reservoir cause ink to flow more freely. Whilst irritating, however, it acts as a pleasing reminder of the slight physicality of this endeavour, connecting me with the paper in a way that entering the words onto a keyboard could never achieve. The Visconti's showy exterior seemed more than fitting to accompany a hugely expensive bottle.

However, as Diageo evidently did not appreciate, my hand-writing is not very tidy, so this was an intimidating task at a physical level, let alone when considering the responsibility that came with it and the fact that my scrawl would be compared by the eventual owner with those contributed by Dave Broom ('a Glaswegian who gets paid to drink and then write and talk about it'); Charles MacLean ('Scotland's leading whisky expert' apparently) and Georgie Crawford, the distillery manager.

I decided after some thought not to produce the requested tasting notes, but to write a letter for the buyer, and this is what I said (if it was you, you can skip this bit as you've presumably already read it).

Poetry, wrote William Wordsworth, is 'emotion recollected in tranquillity', and this Lagavulin poses the interesting question of whether or not whisky may approach the condition of poetry. Having invested so much in this bottle, both wisely and generously, may I add, your emotions on first tasting it will inevitably be both varied and complex, as is only right. But quite possibly, as the experienced taster of whisky that I take you to be, you have no need of my opinions and impressions. Perhaps, to borrow shamelessly from another Romantic poet, round many Western islands you have been and much have you travelled in the realms of gold. And yet ...

And yet, surely we have here a new planet to explore — for this may just be the tastiest Lagavulin ever. However, I write as one who, broadly speaking, remains to be fully converted to the charms of much Islay whisky, finding it, on occasion, lacking in grace and subtlety. However ...

Here, however, we breathe its pure serene and encounter a fully-matured spirit of exceptional quality and distinction.

Put peat fires and smoke to one side: they're present, of course, but it's facile not to go deeper. The first impression on the nose is of the barley sugar twist of my childhood and candied peel — the very apotheosis of dried fruit. While the palate is, at bottling strength, firm and assertive, there is an underlying serenity to the taste as it develops and slowly evolves. This is a majestic and complex whisky, layered, sweet then drying, full of the whispers of history, finishing with fine salted bitter chocolate mints.

Drink, as you must; share, as you should, with pleasure and pride in the knowledge that all our knowledge of Islay will grow as a result of your benevolence and generosity!

Slàinte Mhath!

I have no idea if the buyer appreciates Keats, or even knows who he was, but I do know that bottle number one received a winning bid of £7,300. Lagavulin owner Diageo donated the entire proceeds to Islay Heritage, which, combined with the donation from the auctioneers of their buyer's commission, brings the total raised from the sale to £8,395 from this bottle alone. The buyer, as it turns out, was a German gentleman who has requested anonymity. He did, however, volunteer the information that he would happily have paid more.

Today, Lagavulin is in splendid order, having been preened and polished for the many visitors who passed through its doors in 2016. The legacy of the 200th anniversary is some very fine whiskies, some happy experiences for guests and a permanent contribution to public understanding and appreciation of the island.

I, for one, will certainly appreciate walking out to Dunyvaig Castle and learning more about it. At present it is little more than ruined walls and scattered stones. But they have stories to tell and every visitor will be just that little bit richer for hearing them. Well done, Lagavulin and Diageo.

Back in the distillery's cosy, tongue-and-groove panelled visitor centre there is much to see. Despite its humble presentation, most visitors will gaze reverently at the last sample of new-make spirit from Malt Mill in its simple, even underwhelming glass case, in a modest bottle with a very scruffy handwritten label. 'Malt Mill. Last Filling. June 1962', it records, laconically.

However, to my surprise, it seems that the bottle has only been there since July 2012, prior to which it was in the personal possession of the Lagavulin distillery managers, each of whom appear to have passed it on faithfully to their successor. Today's manager, Georgie Crawford, who played such a part in the anniversary year, thought it best that it was on public display so brought it out of hiding and back on view. Now, if only someone could discover a bottle of Malt Mill—why, you could almost make a film out of that story.

Also on display is an imposing black and white portrait of a fierce-looking gentleman in full Highland dress. It's easily overlooked, which is a shame, for this is Sir Peter Mackie—'Restless Peter' as he was known, for his energy, enthusiasms and egotism. I get the distinct impression he was, on occasion, something of a handful—as mad as a box of frogs, in fact.

The Mackie family connection with Lagavulin began in 1850 when James Mackie, a Glasgow spirits merchant, became a partner. By 1861/62, he had come to dominate the partnership but it was not until 1878 that his nephew Peter joined the business. He was one of the great pioneers and entrepreneurs of the early twentieth century whisky industry, also eventually owning Craigellachie distillery on Speyside and Hazelburn in Campbeltown, and creating the White Horse blend, which is sold to this day. Lots of it is drunk in Russia where it remains a popular buy in that most volatile of markets.

But Lagavulin seems to have been his first love. It was he—on losing the sales agency for Laphroaig—who set up the Malt Mill distillery in the ultimately futile attempt to distil his own version of the neighbouring whisky. Mind you, he only did that after his court action to retain the agency had been dismissed; his attempt to block Laphroaig's water supply had been foiled and he tried to buy all the land above the distillery! Not a man to be trifled with.

As related earlier, Malt Mill came and went. Restless Peter never made a Laphroaig, despite hiring some of their staff and copying the still design. But he used the whisky, of course, notably in a blend called Mackie's Ancient Scotch. Bottles of that turn up from time to time, fetching ever more fantastic prices on auction sites.

There is a whole book to be written on Sir Peter Mackie. His verdict on David Lloyd George, following the notorious 'People's

Budget' of April 1909 which saw a penal increase in the duty on spirits, has passed into history: 'The whole framing of the Budget is that of a faddist and a crank and not a statesman. But what can one expect of a Welsh country solicitor being placed, without any commercial training, as Chancellor of the Exchequer in a large country like this?'

Restless Peter was not impressed, but he did hire Alfred Barnard to write a pamphlet for him and much may be forgiven of those far-sighted and benevolent whisky entrepreneurs who commission indigent whisky writers. The booklet, one of six that Barnard produced for various distillers, is now a rarity, though it has been reproduced in facsimile.

In it, as might be expected, Barnard eulogises his client's product. Mr Mackie, we learn, uses 'the finest barley procurable'. Every distillery you ever read about does this, by the way; I'm left wondering who uses all the cheap stuff, but then if you think about it, in a poor season you could reasonably purchase a poor quality crop and still describe it as the finest procurable — it may well be that the best available isn't very good, but it's still true to say it was the finest to be had.

There are, says Barnard, 'no peat mosses in Scotland like those of Islay'. Well, there wouldn't be would there. On the peat bogs themselves, Barnard writes lyrically of the 'pleasant hours [spent] in sauntering over the mountain slopes' and observes, apparently without irony, the pleasure to be derived from 'watching the groups of diggers and carrying-women laden with baskets of the rich brown turf'. Indeed, I too like nothing better than watching other folk work hard in the sure and certain knowledge that I don't have to spend my life cutting peat and manhandling it across a bog. I've tried cutting peat and it's not easy, and I daresay it doesn't pay very well. Possibly better than writing though.

It's easy enough to poke fun at Barnard (not that it pays any better), but there is much of interest in the little pamphlet. For one thing he maintains that Islay whisky is best made with malt that has been prepared over dry peats rather than damp, which is, broadly speaking, the opposite of current practice where damp peat is used to encourage the production of smoke and maximise the efficient use of an environmentally sensitive product.

He also goes on to describe a ten-year-old Lagavulin 'matured in a sherry cask, very mellow and full-flavoured, which possessed an exquisite aromatic odour'. This is as detailed a tasting note as you will find anywhere in Barnard, who did not run to the highly coloured and extravagant descriptions that have infected so much whisky writing today. It's good to know that at least two of the 2016 commemorative drams hark back to this 1904 reference to sherry casks.

His final remarks are fascinating. Lagavulin was then making around 110,000 gallons annually (up from the 75,000 gallons he recorded in 1886) and it was apparently 'in such demand that the orders exceed the output'. No change there, then.

Barnard concludes, 'Lagavulin has a high reputation at home and abroad; as a single whisky its reputation is unique, and it is one of the few Highland whiskies that can be drunk alone.'

Mind you, alone or in company, drinking Lagavulin can get you into hot water. The pious remarks of the Rev. Robertson of Kildalton in 1794 are often quoted. 'The island hath the liberty of brewing whisky, without being under the necessity of paying the usual excise duty to government. We have not an excise officer in the whole island. The quantity therefore of whisky made here is very great.'

Nearly 100 years later another Islay Church of Scotland minister, the Rev. Mr McColl, was the unwitting (and doubtless unwilling) source of an incident which the mighty *Peterhead Sentinel and Buchan Journal* described as a 'screaming comedy'. The hapless cleric had fallen out with his congregation who requested that he be removed from his post (this is possible in the Church of Scotland). The grounds for his removal were not some ecclesiastical matter or a dispute on arcane points of doctrine, but a matter altogether more worldly, which drew the ire of the fearsome General Assembly, a notoriously humourless gathering. As the paper reported the case, Lagavulin whisky played a central role.

'Fortune and popular election had made Mr McColl minister of an Islay parish where the supporters of the kirk celebrate all fasts and feasts sacred or secular, with the most powerful whisky that Lagavulin distillery can supply; and his physical training had not fitted him for the post. He himself was not indisposed to conform in every possible way to the custom of the country, and when the occasion demanded that an eighteen-gallon cask of whisky should be consumed in three days he took his fair share with the rest of the congregation in the performance. But everyone is not made with the same capacity as an Islay elder, and Mr McColl's form was not equal to that of such men as the witness who could drink three bottles at a sitting and "not be hindered" from his work. He did his best to compete with these heroes, but in vain; and they, for their part, were ashamed of a minister whose head was so much weaker than the Islay standard. The three-bottle men, therefore, complained of him to his ecclesiastical superiors, and Mr McColl had to go, as a warning that Islay ministers must be of equal drinking capacity with their parishioners.'

I cannot help feeling that the Rev. Mr McColl would have been happier preaching in Dorset!

However, enough, for now, of Lagavulin. It's time to drop into one of its neighbours—possibly Laphroaig to the south or, just a mile or so along the increasingly twisty A846, Ardbeg, probably the ultimate cult Islay malt.

But actually, before doing that I'd like to take you about four and a half miles beyond Ardbeg (don't worry, we will return) to Kildalton, where there's something interesting to see. The most obvious point of interest is the remarkably well-preserved early Christian cross that still stands in the grounds of the ruined Old Parish Church.

The cross is said to date from the second half of the eighth century and is closely related to three similar crosses on Iona. But there's no distillery there, so I've never been. That's my loss, I realise, but one can't get everywhere and do everything, try though one might.

Everything you want to know about the cross, and probably more, is contained on an interpretation panel at the site or on the web. What isn't mentioned, though, is the habit that appears to have developed in recent years of visitors leaving a small coin at the base of the cross.

When I visited with the charmingly gullible group of American journalists mentioned a few pages ago, I took it into my mind to persuade them that the tradition was that they should remove some of the currency as a souvenir of their visit. To set the ball rolling I picked up a little of the cash and made as if to pocket it; one gentleman followed with alarming alacrity and I wondered what I had started, until he realised his companions were smirking at such naïve cupidity. For some strange reason he didn't seem to find this amusing; they did though, and I still find it droll, if childish. (May I just record here that I put the money back and added a few bawbees to the pile, as did the rest of the party.)

It was undeniably a childish prank, though, and unworthy of the solemnity of the place. Of even greater interest, however, is the tea shop at Kildalton. This consists of a picnic table, some cool boxes containing milk and cakes, two vacuum flasks with hot water and an honesty box. Tea or coffee is £1 and cakes (lemon drizzle, sticky date fruit cake, lime and coconut slice or coffee and walnut traybake) are £2.[1] You make your selection and your own tea or coffee, and put payment in a metal cash box. Change comes from the small box — as a tiny concession to incipient criminality, both are firmly screwed down. Otherwise the entire transaction is conducted on trust.

I have no idea who runs this, bakes the cakes and collects the cash, but they do a very fine job. Except for stock control: there has never on my recent visits been any lemon drizzle cake, my favourite, and if I could find the management I'd deliver a very stiff complaint. It's really too bad. My Americans seemed to find it all charming and eccentrically British and spent a considerable amount of time (more than on the eighth-century cross, I thought) discussing it on Twitter, Facebook and Instagram, to increasing levels of amusement. They didn't buy any cake or coffee, however, this not being a branch of Starbucks and the transaction being entirely beyond their comprehension. Perhaps, like the enthusiastically amateur pipers on the Royal Mile, the Kildalton Tea Table should charge for photography. That would make for an interesting honesty box.

A more conventional food offering is to be found back at Ardbeg in their Kiln Café, so called because it's located in an old maltings where once there stood, I presume, a kiln. This is where Islay's ladies who lunch rub shoulders with bearded bikers and fierce Scandinavians. Quite worryingly, I observed some tattoos of the Ardbeg logo on a couple of the visitors (not the lunching ladies, though). Adherence to a cult doesn't come much stronger than having its identity permanently marked on your body.

This is all quite a recent phenomenon. For all that Ardbeg was

---

1 My editor insists that I qualify this information with the all-important phrase 'at the time of writing', presumably to prevent readers experiencing higher prices at some future date from suing me for the distress caused thereby.

respected in whisky circles, and for all that Barnard recorded it as Islay's largest producer, it had fallen greatly out of favour by 1982 when then owners Allied Distillers closed the distillery and made the eighteen staff redundant. Those were hard times for the whole whisky business, but the impact on Ardbeg as a distillery village which dated back at least 150 years was profound.

I can recall standing in the distillery yard some time in the mid-1980s looking at the site — then dilapidated and forlorn. Michael Jackson visited around the same time and summed it up very well, describing the buildings mouldering away in the damp, unforgiving Hebridean air as in a 'Gothic mood'. A visit was quite a depressing experience — no one appeared to bother you if you wandered around (presumably no one cared very much) and, even on a sunny day, the place seemed desolate and cold, on the verge of dereliction and surely destined for permanent closure. I guess that, like Tobermory, there is a sense that it survived this period simply because there was no other use for the buildings and it was too expensive to demolish them.

As late as 1989, Jackson could write of Ardbeg that it 'has not operated since 1983, and its future must be in further doubt...' Indeed it was, and that reminds us that it's not long since this style of peated whisky was unfashionable; close to unsaleable, even. Blenders only required tiny quantities and the single malt market, such as existed then, ignored it, all but for a few determined, hard-core enthusiasts.

Michael Jackson was a dedicated true believer though and wrote with feeling to defend the style. So too Jim Murray, another whisky writer, who commented on the distillery's bleak future as late as 1992 in The Scotsman, and was energetic in promoting Ardbeg. In Soho, the retailer and writer Wallace Milroy made sure it was always in stock in the shop he ran with his brother, Milroy's of Soho, for many years a place of pilgrimage for whisky lovers.

Things change though. Tastes change. Corporate policies change. Even companies wedded to their old worn-out blends eventually see the future in island malts. Had you told me in the mid-1980s that Ardbeg would come to life again, that I would find the car park overflowing, the buildings glowing with life, a busy café filled to bursting with an eager, hungry queue at the door, the distillery on the verge of full production and every new bottle released sold out within days, sometimes hours—well, I would have laughed bitterly. So would you.

Yet that is today's reality. In an irreligious age, only a few miles from the grey-green epidioritic[2] certainty of Kildalton, new gods are worshipped. Hardcore devotees bear their tattooed logos as a badge of pride, of community, of belonging.

This has been Ardbeg's crowning achievement for, despite all appearances and the slick marketing, this is not a renegade independent, hurling defiance at the multinational giants. Behind Ardbeg stands first Glenmorangie plc, who brought it back to life in 1997, and today the muscle of the French group Louis Vuitton Moët Hennessy (LVMH) a global giant of luxury marketing. The admittedly delightful and quirky brand stories are carefully crafted artefacts, each curated to perpetuate an image of the plucky islanders as audacious survivors, living on and by their wits, quirky inheritors of an all-but-lost tradition.

It's a page from the Bruichladdich playbook. Ironically, they wrote the script first, but Ardbeg delivered it with more polish and élan. Where Bruichladdich provoked, Ardbeg charmed. It is ironic that both have ended in the corporate clutches of rival French giants.

---

2 The correct geological term for the stone from which the Kildalton Cross has been fashioned, used here adjectivally, possibly for the first and only time, to emphasise the rock-like certainty of the early Christians who carved it—their faith was, it will be remembered, built upon a rock. (See Isaiah 28:16, also Matthew 7:24 etc.)

Both have skilfully recognised how their island reality fits the zeitgeist. Ardbeg represents a prelapsarian world; whisky's Arcadia, set in an idyllic, pristine landscape apparently far from everyday concerns. It's an escape, for a day or a week; a dram-tastic fantasy made manifest, with layers of peaty complexity to engage the senses of even the most dedicated 'smokeheads'.

But the revival had to begin somewhere and, if Murray, Jackson and Milroy championed the brand in the UK, that would have meant nothing without Glenmorangie's purchase and subsequent investments. Within the industry, it's said that by the mid-1990s, there were several suitors interested in acquiring the distillery. As Allied Distillers then also owned Laphroaig and saw that as fulfilling all their requirements for blending and also for a single malt brand, they finally acknowledged a willingness to relinquish control. But not to another large rival who would threaten Allied's position—the new owner had to offer security for the distillery without overt competition. So it's widely believed that Glenmorangie's offer wasn't the highest. Allied could, it seems, have squeezed a little more out of the old place before waving goodbye. But in the quixotically sentimental way of the drinks industry, Allied took the £7.7 million offer for the distillery and what little stock remained partly because they liked the image of benevolence, partly to thwart major competitors and partly because they genuinely believed that Glenmorangie—then independent and increasingly focused on single malts (what a singular idea)—represented the best future for Ardbeg.

Though Allied Distillers were not to survive long as a corporate entity, being eventually absorbed into Chivas Brothers (themselves a subsidiary of Pernod Ricard), their decision on Ardbeg was to prove a far-sighted one. Glenmorangie immediately had to spend money here, on essential upgrading and refurbishment, just to get some production restarted. More than a quarter of a million pounds was spent but, conscious of the need to generate revenue in short order with little depth of stock, nearly three times that was required to convert one of the old malt kilns to a café, shop and visitor centre. Further expenditure on plant, equipment and the site landscaping and car parking has continued to flow freely, and the whole location is now virtually unrecognisable from just twenty years ago.

So we should salute Glenmorangie's work here. Less than thirty years ago you could hardly give this whisky away, and although it has all turned out wonderfully well for all concerned, that could hardly have been guaranteed when they parted with that cheque and inherited the run-down site. Things may have been looking better in 1997 than they had done for some years but this was still something of a gamble, even if it now looks a brilliant strategic move.

Now, there has been something of a tendency in recent years for whisky marketing to concentrate on individuals, branding them as heroic figures and turning the honest, hard-working but low-profile distillery manager of the past into a sort of rock star. It's publicity, of course, and some live up to it, but some don't enjoy the adulation and want to step out of the spotlight. A few others have their heads turned by it, sometimes to unfortunate effect forgetting that their 'iconic' status is a deliberately crafted image and reliant upon the efforts of many others with a lower profile, not least a PR person somewhere.

For some reason, this phenomenon has been particularly prevalent on Islay. It's a combination of the fans' need to believe, the ebullient and extrovert personalities of one or two key characters on Islay and some deft marketing. There's also the human factor, a reaction to the consolidation of the industry over the past three decades. Providing a name and face for consumers to focus on glosses over the scale of the businesses behind these comforting features and, for brands like Bruichladdich and Ardbeg, allows them to portray themselves as resisting the tide of corporate uniformity, even as they are sucked into the giant's embrace.

Personally, I've tried to resist the easy lure of identifying hero figures and describing individuals, however talented, energetic and visible, as 'legends'. It doesn't feel very Scottish, for one thing, and it omits the contributions of all the people left in the shadows. Making whisky is a team effort and for every celebrity distiller or blender there's an unsung accountant or salesperson in some hot, dusty land whose contribution passes without due comment.

Magazines and blogs make it worse with their awards. Distiller of the Year. Visitor Centre Manager of the Year. Brand Ambassador of the Year. The roll call rolls on; publicity is generated; dinner tickets are sold; and trophies presented. Next year the caravan moves on — are we any the wiser? The 'icons' and 'legends' that we read about in such breathless prose are merely individuals doing their job — well, undoubtedly, but they are mortal like the rest of us and prone to the same stresses, temptations and flaws that mark us all as human. Just because your job is perceived as desirable and you do it in an agreeable place (well, perhaps not in December's gales when the planes and ferries have been cancelled for several days running and the supermarket has run out of dulce de membrillo[3] to go with the last piece of manchego — those #MiddleClassWoes, even on Islay, are so very trying) doesn't make you superhuman.

And don't start me, please, on 'passion'. This useful little word is now grossly over-used, a lazy substitute for thought, hollowed out by advertising's large promise, sucked dry by PR hucksters and cynically manipulated by idle copywriters, anxious for their lunch. If I had a pound for every time it has popped unbidden into my inbox I could have retired rich. I loathe it with a fierce and pure

---

3 Quince paste jelly, don't you know, a standard of North London kitchen suppers. It has even crept into tasting notes, God help us.

and virtuous intensity and have resolved never to use it. Never.

And yet... and yet. This is Ardbeg we are speaking of; all normal rules seem to be suspended. Lots of people worked to bring Ardbeg back to life, and lots of people continue to live and work there. But if one person can represent the place, it's the Ardbeg Visitor Centre manager, Jackie Thomson, who came here to a building site in 1997 and has never left. She fell in love with and adopted Ardbeg, cherishes it, exemplifies it perhaps even more than the production team themselves (she'd probably deny that).

It hasn't all been plain sailing and there have been personal costs, yet it's now impossible to think of Ardbeg without calling Jackie to mind. In fact, I have to say it: her passion for Ardbeg transcends any job descriptions and she is known by all who visit here. Curiously, she is not an Ileach and in fact moved from Elgin, but the island has cast a spell on her and, in return, she seems to have worked some kind of magic there.

However, for all the attention Ardbeg receives in the whisky press, on social media and from more than 120,000 enthusiastic members of their 'Committee' (an online community receiving marketing communications from the distillery), the fact remains that this is quite a small distillery, even by Islay standards.

Currently, production runs at around 1.2 million litres annually, though they will admit to trying to 'squeeze a wee bit more than 110%' of that in order to meet the anticipated future demand. By working 24/7 I was assured that 2016/17 production might hit 1.4 million litres. Just as a contrast, Caol Ila makes around 6.5 million litres and Bunnahabhain is capable of some 3.4 million.

While that is going on, there have been changes to the entrance and visitor car park, which I'm sorry to say feels a little too corporate and sanitised now; the water supply and malt intake have been improved and the old car park turned into an open-air public 'event space'.

They do tend, I feel, to take themselves terribly seriously at Ardbeg and there is something of a messianic tone to some of their communications. Their constant attempt to seem funky, and their overly defensive need to play the underdog can grate after a while and, it may be just me, but there seems a sanctimonious feel to some of the web copy. 'A whisky that's worshipped around the world'. Please.

It works for their followers, however. Such is the enthusiasm of the acolytes that recent releases have sold out in short order. This was certainly not the case when Messrs Jackson, Murray and Milroy were such powerful advocates and, had you bought a few bottles back then and cellared them with care, you would have seen a fantastic investment return — not that I would recommend this. In the case of limited expressions such as Alligator, Ardbeg Day and Galileo, those retailers fortunate enough to receive any stock at all were able to pre-sell it before the bottles could even reach their shelves. Today these bottles fetch three-figure prices on whisky auction sites.

Better we don't question whether these will be drunk or simply hoarded for 'investment' but note that collectors of miniatures are even more frenzied in their pursuit of these rare releases. Here's an anecdote to illustrate just that point.

On its release, the Ardbeg people thoughtfully sent me one of just 450 Galileo minis as a sample, along with a tiny, rocket-shaped cocktail shaker (too small to actually use, it was really just a novelty — journalists are easily amused, or so PR folk imagine and, of course, they have darker motives for such apparent munificence). Very foolishly I mentioned it to a mini-collecting acquaintance who, just before he began frothing uncontrollably at the mouth, begged me to part with it. I explained that I couldn't as I planned to drink it for tasting notes for an article I was writing. There was a noise, after which he may have fainted.

But numbers were mentioned—scary numbers—numbers several times larger than my perfectly reasonable fee for writing the article, resulting in something of a moral dilemma. Who would ever know? But I issued a challenge: get me a regular bottle and the mini can be yours. Within forty-eight hours he had pulled it off.

Quandary resolved, I tasted the whisky and wrote an honest story; Ardbeg got their publicity and the magazine their article; I was left with most of a bottle; and the mini and its little rocket went off to a place of honour in a collector's cabinet. Happiness all round. All's well that ends well you'd think.

Well, no. Shortly afterwards I saw one of these little bottles and its companion sell for £940 on a whisky auction site. There was some controversy on social media, with commentators suggesting that Ardbeg were corrupting the Fifth Estate to get publicity. They no longer send out these branded minis. At least, not to me they don't.

I mentioned that it was, relatively speaking, a small distillery. In fact it has just one pair of stills, with the spirit still requiring two wash still runs to be charged. That's not the most interesting thing though: if you want to impress your friends when you get to the still house (and who wouldn't want to earn their 'whisky know-all' badge), direct their attention to the curious additional pipework on the spirit still, and point out that this is a 'purifier'.

You can explain that this device, unique on Islay, but not unknown elsewhere on Scotch whisky islands (e.g. at Talisker, Harris and Scapa), is designed to return heavier vapours back to the still for further distillation. Technically, it's referred to as 'reflux', the idea being that it promotes a gentler, more refined spirit, perhaps fruitier and more complex. Here it's asserted that the purifier contributes to Ardbeg's signature balance of assertive peat smoke and tempering fruity sweetness.

Also said to contribute to the flavour is the fact that all of

Ardbeg's production is matured on this island (like Kilchoman and Bruichladdich in this regard). This speaks to the great 'terroir' debate, which raises such passions in whisky circles. It does mean, however, that there is a lot to see on site, which presents a nicely integrated view of distilling and is really one of the more charming Islay distilleries.

While plenty of cash has evidently been spent, the buildings retain an old-world charm and the distillery layout is appropriately labyrinthine. It's very pleasing to walk through and over the old malt bins, and inspect the model maltings, old pictures and disconnected spirit safe that are on show to visitors. The still room is hot and generally crowded, as somehow you want it to be, and the babel of foreign languages and accents that are to be encountered in the shop and café is a curiously satisfying testament to the global appeal of island whiskies, especially those with such a romantic story.

If Ardbeg is really calling you to Islay, then it is possible to stay at the distillery's Seaview Cottage, the former manager's house (he lives in a cave in the nearby rocks now).[4] For £250 a night, up to six of you can stay in considerable, self-catering comfort in modish designer surroundings.

In fact, I was so impressed with one feature here that I ordered one of my own. This is the magnificent and very, very efficient Hotpod wood-burning stove (see hotpod.co.uk). Actually, this is not a stove—apparently, it's an exothermic oxidizing reactor. Made out of re-melted brake discs, train wheels, old engine parts and so on this is, I promise you, the finest and best-looking wood-burning stove you can buy—and I'm not even on commission. Ours gets admiring glances from visitors and some people have even taken videos of it. However, be warned, it does get very hot.

---

4 Before you feel sorry for him, I actually made this up.

A desirable quality in a stove, but worth watching out for.

The idea of staying at a distillery isn't actually a new one, or unique to Ardbeg. As I've mentioned, it used to be possible to do something similar at Bunnahabhain (not as stylishly) and today Bowmore have a range of cottages available to rent with slightly more conventional décor, but equally as comfortable. They do all book up pretty quickly though, as there is considerable demand.

Ardbeg then is something of a hotspot for whisky tourism — literally where the road ends for some visitors. However, as I say, better to go on and visit the Kildalton Cross Tea Table and take the opportunity to walk on the deserted beaches you'll pass on the way. But we've done that, so back we head to Port Ellen, but not before dropping in on another double-centurion distillery, Laphroaig, which celebrated its 200th anniversary in 2015.

Depending on your direction of travel it is the first or last distillery in this group of three. Like all these Islay distilleries it has a chequered history and, in all probability, there was distilling carried out here prior to the first licensed operation in 1815. Today it's part of the giant Beam Suntory group and therefore under the same ownership as Bowmore.

I hadn't visited the distillery in some years, so, on calling in, was disappointed to find it in silent season. A large number of visitors, vaguely unsure why they were there, were milling about and nothing much seemed to be happening. There were some staff, but they were engaged in the shop so I wandered off into the little museum, which was cool, quiet and restful despite the hordes outside.

Two things did put me off the Laphroaig site, though. The first was the signage which, though well designed and presented, had obviously been written by an American, with American spelling. A small and perhaps insignificant thing, you may feel, but it did grate on me and seemed culturally indifferent and unsympathetic to the place.

But even before I got to the signage, I was disconcerted by the display of tasting notes that had been used to decorate the wall of the walkway into the distillery. Laphroaig had been running a consumer competition, asking drinkers to submit their own tasting notes, the 'best' of which would be displayed at the distillery and on a website where they were collected under the headline 'Opinions Welcome'.

You might think Dick Francis' 'smoky, peaty, oak-aged historic Laphroaig' to be adequate. The brand's fans did not and evidently sought to outdo themselves with ever more extravagant descriptions. The website, especially the linked short video, is fun, but somehow I found the panels on display a little contrived and they seemed to confuse rather than inform. It's certainly more orientated to a US audience than the UK or European consumer, but, having said that, I cannot but admire any brand honest enough to feature a tasting video in which consumers grimace, pull faces, associate the product with school lavatories (not, I assume, in a good way), seagull's armpits (implausible, but that's what the chap says) and so on. Of course, in the end, most of them love it and the point is that, like its neighbours, it splits opinion, but this is well done.

Much is made at the museum, and in the distillery's literature, of how one-time owner Ian Hunter passed on the distillery to the celebrated Bessie Williamson, one of the few women to own and operate a distillery in the modern era.

Hunter may not have been the easiest of employers; even the publications approved by the distillery describe him as 'meticulous about everything and very suspicious'. No outsider was allowed into the distillery and he refused permission for newspaper coverage or photography, presumably as a result of the long-standing dispute with Peter Mackie. One wonders what he would make of today's visitor centre and website.

Though it seems strange that any writer would endorse the

suppression of another's work, there is an approving reference in some of the recent literature about Laphroaig to Hunter attempting legal action in the early 1930s to block the publication of a book, on the grounds that it gave a detailed description of the distillery. It is said to have been written by one James Whittaker, described as a cooper.

That there was a legal action seems clear, but the story appears to have been distorted over time. The 1915 County Valuation Rolls record a James Whittaker, described as a Cooper, resident at Laphroaig, but it was not him who wrote the book.

It was, in fact, his son, also known as James Whittaker and the book in question appears to be I, James Whittaker, which was published in London in 1934. It was favourably reviewed, and must have enjoyed some interest in that the foreword was contributed by Gilbert Frankau, then a minor literary celebrity, though today he is largely forgotten and his works obscure. George Bernard Shaw is also believed to have possessed a copy and, with his Fabian beliefs, will have been sympathetic to the contents.

The chapter on life at Laphroaig (Whittaker calls it 'Lafroyayg') is revealing of social conditions and attitudes, but it seems to me unlikely that Whittaker can have been writing about the distillery in any technical sense as he is describing his life there as a child during the First World War. We can never know exactly what was suppressed, though as Whittaker noted much later that 'the really shocking stuff was carefully excised' (presumably as a result of the legal action; he claims the book is 'as true as the law permits'), it must have been controversial indeed. Perhaps the elder Whittaker was less than flattering in relating to his son the distillery's treatment of employees. Hunter surely will have objected to the vivid descriptions of the living conditions of the tenants at Laphroaig, especially as his family controlled the leases of all the properties and was their direct landlord.

It's easy today, in our twenty-first-century comfort, to gaze at the black and white photographs in the Museum of Islay Life and elsewhere as if they were exhibits in some anthropological display, forgetting that they show real people living just 100 years ago. Islay today is frequently described as some sort of island idyll, a bucolic paradise, but life as recorded by James Whittaker was very different.

He writes with raw feeling that Laphroaig was 'a filthy, miserable place. There were no proper roads or paths, mere tracks through the bog mud led from door to door. The houses were old and dilapidated, most of them in bad repair; some were crazily thatched and obviously in need of new roofs. The houses stood all over the place: in no spot were there more than two together. They seemed to draw away from one another with sullen mysterious looks'.

There is much in this vein: 'There seemed to be a curse on the place and the people... They were lethargic, dull, hopeless'. Later his family were turned out of their modest accommodation, presumably by Hunter's father, and required to live in 'a miserable mud hovel' from where Whittaker relates he was obliged to 'hobble in agony to the school at Port Ellen, miles away, through the snow and ice-covered roads, and the rough road-metal split and cut my blue-purple toes many times'.

Reading Whittaker's autobiographical memoir is an antidote to the current romantic narrative on Islay. 'Lafroyayg', he writes 'was like a loathsome disease which silently and stealthily creeps upon one, marring, disfiguring, stifling and numbing the soul'.

At this distance, we can see Whittaker's book as part of a genre that flourished before the war, commenting critically and from a left-wing political perspective on the social conditions of the British, largely English, working class. Better known examples include J.B. Priestley's English Journey (1934); two highly influential books by George Orwell, *Down and Out in Paris and London* (1933) and *The Road to Wigan Pier* (1937); and the lesser-known *I Was a Tramp* by John Brown (1934). Perhaps the major difference between Whittaker's work and those of Priestley and Orwell, apart from their respective literary merits, is that the latter two authors were members of the intelligentsia, observing the condition of the poor (albeit at first hand) from comfortable middle-class backgrounds and an intellectual and political standpoint, while Whittaker had direct personal experience of a squalid life in the slums from which he had escaped with some difficulty.

'Lafroyayg means three things to me: poverty, misery and loneliness'. We may safely conclude that Ian Hunter didn't care for that description at all, and it is perhaps little surprise that he tried to suppress this unwelcome and subversive view of 'the beautiful hollow by the broad bay'. Such an image would hardly sit well with his pioneering efforts to promote his brand following the repeal of Prohibition. Laphroaig is said to have enjoyed a medical exemption and have remained legally available in the USA during Prohibition, providing a sound base for further sales. Publicity such as Whittaker's would have been unfortunate to say the least and Hunter will doubtless have been gratified that *I, James Whittaker* was not a great commercial success, though it was reviewed sympathetically enough. I'm glad to have the opportunity to set the record straight and restore something of its lost reputation.

But Hunter took Laphroaig round the world, expanded the distillery and was progressive enough in his outlook to recognise the remarkable contribution of Bessie Williamson, who took over the distillery in 1954. She's certainly one of whisky's great characters, eventually selling Laphroaig to Long John International, after which it passed to Allied Domecq (who then also owned Ardbeg) and thus to Fortune Brands, Beam Inc. and, today, Beam Suntory where presumably it will rest until the next great corporate reshuffle.

Through all of this the distillery has managed to maintain a distinct identity (albeit with some American signs), with floor maltings and 100% of its spirit matured on Islay. They also pioneered relationship marketing, with the clever idea of granting a lifetime lease of one square foot of Islay to the Friends of Laphroaig, thus generating a steady flow of visitors to the distillery who come to lay claim to their plot. It is not unknown for visitors to attempt to cut out their square foot and take it home, such is the loyalty that the scheme has generated.

One enthusiast who presumably would never go quite that far is HRH The Prince of Wales (or the Duke of Rothesay as he is more correctly styled in Scotland). Prince Charles granted his Royal Warrant to Laphroaig in 1994, declaring it his favourite whisky. One wonders if in time the name will be changed to Royal Laphroaig. After all, it has worked for Royal Lochnagar (Queen Victoria) and Royal Brackla (Queen Victoria and King William IV, the only distillery with the distinction of two warrants), though it cannot be said to have done much for Glenury Royal, which closed permanently in 1985. Such a republican fate seems unlikely for Laphroaig.

If you care for distinctively peaty whisky, Laphroaig strikes me as pretty good value, with the standard ten-year-old currently retailing at under £40 and the more intense, accelerated-aged Quarter Cask only a few pounds more. As the tasting opinions make abundantly clear, not everyone will care for the taste and a little goes a long way, but that only increases the desirability for the enthusiast and makes it even better value for money.

The location can hardly be faulted for romantic drama. These three coastal distilleries are all best seen and approached by sea, a route which makes more sense in the Hebrides than road transport.

What is more, the sea approach reveals some remarkable locations: Dunyvaig, of course, but also unexpected sights such as the old sheep droving jetty at Glas Uig, a small and sheltered bay near to Kildalton. Droving tells us something of Islay's history but the bay's main claim to fame is that during both World Wars German U-boats were known to anchor quietly here and put men ashore in search of fresh water and meat (more bad news for the ovine inhabitants). However improbable the stories may seem today, they are in fact well documented and more than one German submarine commander is known to have returned to Islay to investigate their furtive anchorage in happier times.

I once had the considerable pleasure of seeing the spot from the sea while travelling from Lagavulin to Port Ellen by fast RIB and realising just how difficult it would be to spot a partially submerged U-boat once it was tied up to the jetty. The implausible tale was immediately credible.

Of course, the road is how most of us find our way to these distilleries, but it cannot be denied that for sheer impact the experience of arrival by sea can hardly be rivalled. Having said that, I once travelled to Talisker by helicopter, stopping off for lunch at the Three Chimneys, and that had its attractions. However, expecting to cause a mild sensation on our arrival, we were more than a little chastened to find our machine the third in the helicopter park—yes, there was a 'helicopter park'—and merely a noisy irritant to fellow diners.

Anyway, for all its charms, I did not linger long at Laphroaig on this occasion and continued on to Port Ellen, passing the site of the proposed Farkin distillery. I always think that Port Ellen town presents an unfortunately drab face to the world, though the rebuilding and re-opening of the Islay Hotel has helped.

I've enjoyed two meals here, though I use the word advisedly. The first was a dinner, accompanied by my wife, shortly after the place re-opened. She had stayed there as a child and was interested to see what had been done. The dinner was a memorable experience, so transcendentally dreadful in every respect that, as the shambolic evening rolled chaotically on, we began to speculate on what could possibly go wrong next, behaving on the assumption that we were the victims of an enormous set-up being recorded for some ghastly TV show. The couple at the next table entered into the spirit of the occasion and by the end of a truly appalling evening we were thoroughly enjoying ourselves, a sensation which lasted until the bill arrived and the cold recognition set in that we were actually being asked to pay for what was otherwise a training day for new staff, possibly on day release from a young offender's institution.

With that fresh in my memory I was apprehensive about a second visit for lunch, about a year later, so it is only fair to record that everything (this even includes the furniture and crockery) had improved to the point where the meal was perfectly fine. Bland, unmemorable, but quite lacking in any *Fawlty Towers* component whatsoever.

The occasion was a visit to the Port Ellen Maltings and what remains of the distillery. That is simply disposed of: basically, there is nothing to see. The kiln, mash house, tun room and still house have all been demolished and apart from some dunnage warehouses, some nondescript business units occupy the rest of the buildings. The owners, Diageo, remain bafflingly cryptic when asked about the remaining stock of whisky, which they eke out every year as part of their Special Release programme at ever higher prices—currently £2,500 for a bottle of the sixteenth release. Back in 2001, the First Release was thought overpriced at £110 though today that particular bottle carries a £3,500 valuation.

But, if £2,500 for a bottle seems excessive, there are online retailers who will supply a 3cl sample—marginally more than a single pub measure so you'll need a magnifying glass — for £218.33, including delivery. However, given that the distillery closed in 1983, when it had been producing well below its maximum capacity for some years, there cannot be very much left. Quite what will happen when it's all gone I cannot imagine. They may declare a day of national mourning across Scandinavia, where these whiskies are especially popular.

The warehouses look to be in pretty decent order, which suggests that they contain something of value, with crisply white-washed walls and the Port Ellen name standing out in black lettering to the seaward side. You can walk onto the shingle beach and we did that, stopping for a dram of Port Ellen in the sunshine. It tasted like ashes.

However long the supplies last, Port Ellen will always have a small part in whisky's history for it is here, in 1824, that the first spirit safes were tested to ensure that the equipment had no adverse effect on the quality of the whisky passing through it. Quite why anyone imagined that it would I cannot conceive, but I suppose the precautionary principle holds here. At about the same time Robert Stein and Aeneas Coffey were working on designs for the column still that would revolutionise whisky through the efficient production of grain whisky, the key to successful blending.

Some years later, Donald Osborne, then General Manager, provided special help to one J. Scarisbrick of the Inland Revenue on 'spirit testing in the USA', which formed part of Scarisbrick's 1898 manual on 'Spirit Assaying', which he dedicated as from Port Ellen. Quite why Port Ellen's Manager should be familiar with American practice in this field isn't quite clear, but it is testimony to the reach and influence of this iconic distillery, which was certainly exporting whisky to the USA in the nineteenth century.

The original distillery was shut down in 1930 though, and remained silent until it was effectively rebuilt entirely in 1966/67, and it is the closure of that distillery in 1983 that is so loudly lamented today. But it wasn't closed by accident: there was little or no demand back then for heavily peated whisky, either from the blenders or the nascent single malt market. What's more, the distillery needed money spent on it and Islay looked pretty down and out. Things change, of course, but this was a rational and understandable decision by hard-headed men of business.

But things do change, and Port Ellen is as representative an example of that particular truism as you can find anywhere in whisky. I spoke to the people at The Whisky Exchange, and they estimated that between 2007 and 2014 they had handled more than 400 different bottlings of Port Ellen. Some of that is because merchants are cashing in and some because independent bottlers holding casks realise that their carefully hoarded stocks are in danger of getting over-aged and excessively woody. Eventually there will be no more; the flood will dry up and Port Ellen will join Port Charlotte, Malt Mill and Octomore on the roll call of legendary but lost Islay distilleries.

However, there remains much for the whisky enthusiast to see in Port Ellen in the form of the giant Port Ellen maltings, whose slab-sided buildings dominate the road into the town and overshadow the remains of the old distillery. Until relatively recently, the plant was closed to the public, but Diageo have relented in the face of continued interest and now offers tours during the Islay Festival.

Stepping inside the enormous building is a way to start to understand the scale of today's distilling industry, particularly if you have earlier visited the floor maltings at Kilchoman, Bowmore or Laphroaig. In fact, it's pretty well vital to have seen one of those at first hand to appreciate what is going on here, for this is modern industrial drum malting and without this type of technology it would be impossible for Scotch whisky to have achieved the volumes that are sold around the world. Again, in an industry that likes to present a face of unchanging consistency, we see how technological and engineering progress have been vital to the global success of the blended brands that, in turn, ensure the continued existence of so many single malt distilleries.

The maltings were built in 1972/73 and replaced the on-site maltings at Port Ellen, Caol Ila and Lagavulin. Originally, production was reserved for the Diageo distilleries only, but in the mid-1980s, with distilling at a low ebb, the future of the maltings was in some doubt. An imaginative and forward-looking deal was then agreed amongst all the distilleries on Islay (and nearby Jura) to keep the plant open by all agreeing to take some of their malt from Port Ellen maltings, despite this being owned and operated by their huge competitor. This 'Concordat of Islay Distillers' saved the plant and all the associated jobs at a time of real and sustained economic pressure. Sadly, it has now broken down, though some malt is still produced to the various distilleries' individual specifications.

Internally, everything is on an impressive scale: the silos hold 2,040 tonnes of grain at the site, with another buffer stock of 650 tonnes of barley and 30,000 litres of water (the typical domestic bath requires only 180 litres). The seven malting drums, the largest in the UK, each hold the content of two steeps, the weight of which is now sixty-five tonnes as the barley has swelled to a 45% moisture content. The drums rotate every five minutes for eight hours with hot air blowing through them, provided from the kilns below.

Depending on the peating level required, peat from the Castlehill moss will be burnt in the kiln, with a single, heavily peated batch requiring some six tonnes of peat to be burnt over a typical thirty-hour period. Just to be on the safe side, the maltings maintain a stock of some 2,000 tonnes, enough for around one year's production of malt.

Though Port Ellen no longer supplies malt to Bowmore, from the maltings to the distillery is a distance of around ten and a half miles, covered quickly enough these days. Apart from a dramatic left and right turn shortly after leaving Port Ellen, the A846 runs straight and true across the peat moss, past Islay's airport at Glenegedale—home to some of the most officious security staff on the planet, so don't ever tangle with them—and into Bowmore, the principal settlement.

Of course, it's only covered quickly if you don't have to stop for roadworks, as occurred on a recent visit. But that was actually quite restful, once I had decided not to be annoyed and frustrated. A section of the road was being re-laid with new tarmac and no traffic was allowed to pass in either direction. Had anyone bothered to inform vehicles in advance it would have been possible to use the alternative, minor road that runs almost parallel to the A846 a mile or so inland. However, apparently, no one had thought the convenience of the traffic of any importance whatsoever, so I opened the roof of the car to enjoy the sunshine and the buzzing of various unidentified insects and savoured the unusual experience of an Islay traffic jam in true Hebridean style. A few cars ahead a gentleman fired up a camping stove by the side of the road and made tea. Didn't offer me any, unfortunately. Soon enough we were on our way.

Sadly, Barnard found the drive 'one of the most uninteresting' that he had ever experienced, noting that he had 'never travelled by such a dismal and lonely road'. That seems a little harsh, but he was constrained to proceed at no more than four miles an hour and I can well imagine tiring of the view at such a pace. His coachman lacked any sense of urgency and, as he relates, 'some of us walked many a mile, and were yet able to keep ahead of him'.

He had, however, completely misunderstood the Highland temperament, 'continually remonstrating' with his driver and offering 'nips of whisky to induce him to urge his steeds along'. Clearly the driver had worked out that the slower he proceeded, the more whisky he would consume. Had it been me, I should have fed the driver tea, with the incentive of hard liquor dependent on a speedy arrival. Surely then his lethargic nags would have felt the whip!

However, once the indefatigable traveller arrived he made the best use of his time, devoting nearly five full pages of text and two illustrations to his fulsome description of 'the great improvements and alterations in the works'. This is an extensive amount of space for Barnard, around twice as much as he devotes to any other distillery on the island, and involves a lengthy description of every part of the Bowmore operations.

Full disclosure: having passed lightly over most of my fellow author's efforts on Islay, I should note that I have contributed to the literature with my own description of Bowmore, in a larger book[5] I wrote for the parent company, Morrison Bowmore Distillers.

This had a curious genesis. It was commissioned by the then Managing Director, Mike Keiller, and his senior production colleague, Operations Director Andrew Rankin. Both have left the company now, subsequent to the creation of Beam Suntory, but, at the time, Morrison Bowmore Distillers was a largely autonomous subsidiary of Suntory, and Keiller and Rankin were anxious that its story should be told. While they were model clients, it became apparent during my work on the book that the marketing department regarded the project with some suspicion. Over time, it was evident that their attitude was that if anyone was to commission a company history it should be the marketing department, and I was regarded as something of an interloper to be assisted on sufferance (as I evidently had the ear of the MD) but

---

5 But the Distilleries Went On: The Morrison Bowmore Story. Angel's Share, Glasgow. 2015.

not accorded any substantive help. It became quite clear that the project was 'not invented here', with all that that implies.

Events moved fast when the Beam Suntory merger was announced, and while the book was finished during Keiller and Rankin's term, they had left office by the time the printed and bound copies were delivered. On the last occasion that I visited the distillery it was not available in their visitor centre, yet was prominently displayed in the window of the most excellent Roy's Celtic House gift shop, the fine emporium gracing a corner of the town square. My emails on this curious omission remain, as yet, unanswered.

However, having written the chapter on Bowmore and visited the distillery on a number of occasions, I am happily familiar with the interior of what proudly claims to be Islay's oldest distillery. The company date the foundation to 1779, though Bowmore itself was created as a planned village from the late 1760s and there are some commentators who hold that distilling began here almost immediately. However, by the same token, Lagavulin can make some claims to having been established in 1742 and, as it seems few drinkers will be sufficiently impressed by the claims and counter-claims to alter their brand choice, perhaps it is of little more than academic interest.

Like many a Scottish distillery, Bowmore was 're-established' in 1825, following the radical 1822 and 1823 Distillery Acts which boosted legitimate production across Scotland. David Simpson, who founded Bowmore, was noted for his pledge to inform the authorities on 'any person or persons…concerned in this illegal and destructive Traffick' (i.e. illicit distilling, then not unknown on Islay).

Morrison Bowmore Distillers began life in 1951 as a very different entity. The company, first named Stanley P. Morrison Ltd, was a whisky brokerage and for many years, even after they owned distilleries, the trading of bulk stocks of whisky made up a substantial part of their business. Latterly, they were a major supplier to Japan, at a time when that trade was frowned upon by the industry's more established figures. They also had significant Latin American volumes and supplied a number of UK supermarkets with low-cost blends and own label malts.

Bowmore distillery came into their ownership in July 1963, following some very sharp work by Stanley Morrison. At that time, Bowmore did not enjoy the highest reputation, being run-down and with production at very low levels.

The seventeenth of July 1963 was a Wednesday. The newspapers were full of the passing of the Peerage Act, 1963, the legislation by which the 2nd Viscount Stansgate, more familiarly known as the Labour politician Tony Benn, was able to renounce his peerage and stand for election as a Member of Parliament. In the USA, the first protests against the Vietnam War had begun, while here in the UK Frank Ifield's *Confessin' (That I Love You)* had topped the hit parade. And, no doubt entirely indifferent to any of this, in Glasgow Stanley P. Morrison was enjoying a convivial lunch with his stockbroker in the fashionable Malmaison Grill Room of the Grand Central Hotel.

Part way through the meal he overheard a snippet of conversation, which piqued his interest. Straining a little, he soon became curious to learn more. Under discussion, he soon gathered, was the pending sale of Bowmore distillery by the executors of James Grigor, whose company, William Grigor of Inverness, had bought it in early 1950. The intended new owner was to be Destilerías y Crianza (DYD) of Spain which, today, is ironically also a subsidiary of Beam Suntory.

Morrison finished his lunch abruptly and set to work. Contacting his office, he obtained the phone number for James Grigor's widow. Was it true, he enquired urgently, that Bowmore distillery was to be sold?

It seemed that it was: with three sons attending Loretto, Scotland's leading public school, she had many expenses to meet and so a sale was inevitable. In fact, she added, she would be travelling to Glasgow the following day to conclude the transaction.

Using all of his very considerable charm and adding, no doubt, an emotional appeal that the distillery should remain in Scottish hands, he persuaded her to alter her arrangements and come to Glasgow that very day to have dinner. She agreed, and before that meal was over, he'd bought Bowmore distillery, unseen, along with a bonded warehouse in Glasgow. The deal amounted to £117,000 — distillery, bond and stock.

Shaking hands with Mrs Grigor (I like to imagine, over a healthy dram) the deal was done and Bowmore changed hands. The 'due diligence' consisted of Stanley Morrison's own judgement. His partner, James Howat, was on holiday on Arran playing golf and when telephoned the next day to be told of the purchase is said to have responded querously, 'Where's the money to come from?'

'Well, that's up to you,' was apparently Morrison's reply.

But they had bought a run-down and antiquated distillery with only one customer, which was running at less than a tenth of its capacity. Without an order from the blenders Robertson & Baxter, who only needed supplies because their Bunnahabhain was closed for modernisation, Bowmore would not have worked at all during the 1962/63 season.

Hardly an auspicious start.

In another low-profile beginning, around a month later, in August 1963, Bowmore took on a new apprentice in the cooperage. Though no one could know it, James 'Jim' McEwan was later to play a significant role in distilling on Islay. His name appears from time to time in these pages.

But the purchase of Bowmore, just fifty years ago, demonstrates how radically whisky has changed over those decades. A deal done over dinner. A handshake to seal the bargain. And £117,000 for a distillery—the most expensive Bowmore single malts are now £100,000 for a bottle!

Stanley Morrison had no idea of any of this, of course. He was seeking to secure his supplies of whisky; by owning a distillery he was in a more powerful position as a broker and he gained some measure of respectability with the gentleman's club that then controlled the business. Brokers were, I get the impression, more tolerated than liked, though they bought good lunches, drove fast cars and entertained lavishly on the best golf courses. Such things mattered in the whisky world of the 1960s, and though the broker has largely disappeared with the consolidation of the industry, something of their raffish style lingers on.

There was no thought in the purchase of developing Bowmore as a single malt in its own right. Stanley Morrison believed that such a development would put him into competition with his own customers and considered this both an act of bad faith and unlikely to be significantly profitable in any event. Following his death, however, branded sales of Bowmore began to be developed and, though not as forward-looking as other independents, Morrison Bowmore were relatively early pioneers in marketing single malt whisky.

These sales were developed during the late 1970s and 1980s, though this was a challenging decade for the industry in general and Islay in particular. Most of the island's distilleries, Bowmore included, were on short working; others such as Ardbeg and Port Ellen closed entirely and, as we have seen, Port Ellen was never to re-open.

Bowmore survived thanks to bulk sales to Japan, something that was widely criticised but which proved essential to its continued existence, and because it was in the private ownership of people committed to the whisky industry they were able to take a longer-term and more paternalistic view than those firms driven by the shorter-term view of bankers and fund managers. But for all that, these were trying times, and I recall my visits to Islay during the 1980s as being sombre (at least when not throwing stones at shipwrecks, which will lighten the darkest of days).

As trade slowly recovered and malt whisky began to build a following, Bowmore's reputation suffered a setback. This was the infamous 'French Whore Perfume' incident.

Now, I would like to make it immediately clear I am completely unfamiliar with the fragrances associated with the commoner class of French cocotte or, for that matter, the expensive ones, so it is not a nose that I would expect to recognise nor, if I did, publicly acknowledge. However, by the late 1990s a group of experienced and well-regarded whisky enthusiasts were declaring it to be a new and unwelcome aroma in distillery bottlings from a decade or so before.

It was also, more politely, known as the 'flowery water problem', best described as an overpowering aroma of lavender that certain-ly damaged Bowmore's reputation amongst an influential group of critics. However, the distillery's efforts to close down the debate made things worse. Legal threats were issued and much bad feeling ensued as a result of this heavy-handed approach. Today the damage that would follow the inevitable storm on social media can hardly be imagined, but in a more innocent age, eventually the debate subsided, not least because the problem disappeared from subsequent bottlings as quickly and as mysteriously as it first emerged.

It did, quite understandably I think, set Bowmore back in the eyes of single malt purists and this, combined with the fact that Bowmore is less heavily peated than some other Islay malts, may account for its more modest reputation and lack of true cult status.

Not that the distillery has been slow on capitalising on the interest in Islay whiskies, and some releases—notably the Black

Bowmore—have achieved the standing of near legends. The first Black Bowmore was offered in 1993 at a little under £100. One long-serving distillery employee told me ruefully that he had baulked at the £80 staff shop price, and another related that he had bought a bottle and drunk it. Both decisions are now a matter for regret, with this particular release now selling for several thousand pounds on auction sites.

That, however, pales into relative insignificance when compared to the £61,000 paid in October 2013 for a single bottle of forty-eight-year-old (1964) Bowmore at a charity auction organised by the Worshipful Company of Distillers in London. Earlier in the evening I had been asked to pose for a photograph. Thinking to use a suitable prop, I reached for the nearest bottle which, by chance, was the Bowmore, and thus was recorded for posterity holding the most expensive Bowmore ever sold at auction. In retrospect, I should probably have picked another of the many lots, but fortunately I contrived not to drop it. Never again!

Bowmore today seems in tip-top condition. Being so centrally located, it benefits from good visitor numbers and has cleverly developed the distillery houses in Bowmore Square into up-market holiday accommodation. The company also purchased the Harbour Inn hotel in January 2014.

That purchase was dwarfed the following day when Suntory, owners of Morrison Bowmore, announced a $16 billion bid for Beam Inc. of the USA. The deal proceeded without serious opposition and the combined group came into existence as Beam Suntory. The consequence was that Morrison Bowmore Distillers was downgraded from an autonomous business and the brands placed under the control of an operating company based in Madrid. But the distilleries went on, of course, marking another stage in their evolution, and I got a title and a neat ending for my book.

And this, I think, is where I leave Islay, Queen of the Hebrides. A tantalising, frustrating place, half in love with its own image, half of which has been created for it and imposed upon it in the interests of others. I still miss the emptiness, the derelict, rusting farm equipment and the fin-de-siècle atmosphere, but will not deny I enjoy the better food, drinkable coffee and more cosmopolitan atmosphere.

Where it will all end, I have no idea. There have been booms here before, and busts, and the pattern will surely repeat. I hope for your sake that you sell your whisky collection before then and I hope for the industry's sake that someone knows what to do with the many, many warehouses full of peated whiskies if the current fashion for smoky drams ever fades.

Until then, farewell, Islay. My next trip took me to the twin islands of Harris and Lewis, and back in time some five decades.

8

Harris & Lewis

A flagrant breach of the Fisheries (Dynamite) Act, 1877

I HAVE SUCH INTENSE CHILDHOOD MEMORIES OF THESE conjoined islands—of the memorial stone for a dead whaler's faithful dog; of my father's tense and nervous glances as his car was winched into a net and onto a ferry; of fishing with high explosive; of mother's attempts at Gaelic; of abandoned cars and rotting steadings—that the twenty-first-century reality came as a shock.

However, I had to return: there are two distilleries here now, united only in their isolation, but which could not be more different in character and approach if they were on opposite sides of the planet.

The map confirms that Harris and Lewis are one land mass, with the larger, Lewis, lying to the north and Harris making up the smaller, southern portion. In total the island is the largest in the

Hebrides and, in fact, the third largest in British Isles after Great Britain and Ireland themselves. For all that, they remain relatively little known and are only sparsely inhabited.

Visually, the boundary between the two would seem to fall where the land narrows at Tarbert but the administrative boundary actually lies between Lochs Resort and Seaforth. But why does this matter? I hear almost no one enquire.

Well, it matters here because the first distillery that we managed to see on this visit proudly sells a product labelled Spirit of Lewis (it's a new make, so not legally whisky) and styles itself as located on Lewis. In fact, great play is made of this, which is convenient, as the distillery — it's Abhainn Dearg by the way (or Red River in English) — can legitimately distance itself from the more recent Isle of Harris distillery at Tarbert, which likes to promote itself as the first distillery in Harris.

Clearly one island ain't big enough for the both of them.

I imagine that it suits them both to maintain their distance, with one in Lewis and one in Harris. Perhaps that way they imagine no one will make comparisons. Sadly though, it hasn't worked.

But the geographical distinction, though clearly important to them both, is far from the most interesting or curious thing about Abhainn Dearg. This, in fact, may be the most interesting of all Scotland's island distilleries, so read on.

However, first one has to get there. It's possible to fly to Stornoway, but in this case the trip involved the lengthy drive through Skye to Uig and another ferry crossing, this time to Tarbert on Harris. Today this is served by a modern, smooth and sophisticated ship of the Caledonian MacBrayne line.

Not so on the occasion of my first visit, again for a family holiday in the 1960s. Back then, as I persist in boring my sons by telling them, the journey was accomplished on a more utilitarian vessel that required any car to be lifted onto the foredeck from the pier by means of a sling or net, the hazardous procedure being reversed on arrival, all of which I can now see accounted for my father's un-derstandably stressed demeanour. The fact that the vehicle was not his but an early and prized example of the company car doubtless contributed to his anxiety levels. Explaining to your boss that the car was at the bottom of the Minch would have proved challenging.

Once on board—I am persuaded that the vessel was probably the *Claymore*—we children were sternly admonished not to go anywhere near the saloon, which, of course, we immediately sought out. This was inhabited by fierce-looking men and through the clouds of blue smoke we could just make out that they were drinking glasses of whisky with a strangely furious energy, or so it seemed to us, as my mother's disapproval was quite evident. As to their speech, what we caught as we peered sheepishly round the door was presumably the authentic sound of living Gaelic (as opposed to middle-class subsidised Gaelic).

Our sanctuary was the lounge, where we sheltered against the gaze of these supposed ruffians and trembled at the vicious vibration that characterised our ship, whose engine was evidently unbalanced in some way. I was convinced it would sink and this was why so much whisky was being feverishly consumed. As our vessel juddered and rolled violently, it seemed at every moment that the rivets holding the ship together would be shaken loose and we would surely be cast into the uncaring sea. Here's a tip: never travel with a child with an over-active imagination.

Some years later I came across an exhaustive account (275 large-format pages; I believe that qualifies as exhaustive) of all the CalMac ships entitled *The Kingdom of MacBrayne* by Nick Robins and Professor Donald Meek, and this explained that the *Claymore* had been fitted with four-cylinder engines when five- or six-cylinder versions would have ensured much smoother running. They don't say why such economy was adopted.

They describe the *Claymore* as 'distinguished for her excruciating vibration, with its distinctive and unnerving rhythm'. You can say that again. 'Her vibrations would "wind up" to a crescendo,' they relate 'and then sink away to relative calm before recommencing the cycle'.

It was, I think, the relative calm that was most unsettling, for you had just convinced yourself that the rattling and shaking had finished before the vile cacophony would begin again. I can hear and feel it still, though Professor Meek evidently recalls the *Claymore* with great fondness. Perhaps my memory is at fault, and she never plied the route to Harris,[1] but I surely sailed more than once on this most memorable member of the MacBrayne fleet.

The passage today is innocuous by comparison and accomplished with little fuss or drama. The modern boats resemble small cruise liners, with shops, a coffee bar and a perfectly unexceptional self-service café (the old ships specialised in well-hung seagull fried in used diesel oil, with five-day-old chips). There are excellent and remarkably effective stabilisers for rough seas, the usual vapid yet ostensibly caring customer service announcements and a simple health-and-safety-approved procedure for driving on and off. They are bland, unexceptional, efficient—everything you might want for a short ferry crossing, shorn of any seagoing romance whatsoever—and there we shall leave them.

Or perhaps not. It was always father's custom to recite the following short verse on successfully boarding any MacBrayne vessel. It's actually a parody of Psalm 24 in the words of the Scottish Psalter, though not being a religious man I'm not sure if he knew this—I certainly didn't and for some while credited

---

1 Someone will surely write in and prove definitively that the *Claymore* never served this route. To be honest, I don't really care about the literal accuracy—I certainly had experience of the old tub more than once. Perhaps it was to Colonsay or Coll. We went there as well in my childhood and I carry the scars of the passage to this day.

him with these lines, especially as the last line doesn't really scan satisfactorily. Thoughtfully, I was apt to remind everyone in earshot of this.

The earth belongs unto the Lord
And all that it contains,
Except the Western Isles,
And they are David MacBrayne's.

On arrival at Tarbert almost the first thing you see is the new Isle of Harris distillery, but it being the Sabbath, we drove right on by to find our hotel. Respecting long-established local custom, Tarbert was closed in its entirety and even a short walk around the town passed without sighting a local, or even another irreverent visitor. Time was, there would never have been a Sunday ferry but commerce has conquered observance even here.

On Monday, I had arranged to visit Abhainn Dearg, or so at least I thought. The drive from Tarbert to Uig (the Lewis Uig, not the Skye one where we started, obviously) is a long and, for the first-time visitor, alarmingly otherworldly one. Incidentally, if you've ever wondered why Scotland has so many Tarberts, it's because the name is derived from the Gaelic term for an isthmus. Anyway, using the public road it is nearly sixty miles and takes at least an hour and half. A decent helicopter would cover the twenty or so miles of the direct route in about ten minutes—call it fifteen to allow for take-off and landing—and providing all the way a spectacular view of the reason for the extended round trip, namely the large and steep mountains that block the direct approach by road.

The main obstacle is An Cliseam, a 2,621-foot lump of Lewisian gneiss (very tough old granite to you and me), closely resembling a moonscape. People apparently climb this and the linked peaks, accomplishing something known as the Clisham Horseshoe. Personally, I'd strongly recommend the helicopter.

But that would mean missing out the community stores at Uig ('a Tardis, small and compact on the outside, but inside there is a wealth of produce', and I can confirm this to be so) and the splendid Uig community centre with its tea room, fire station and little museum, all next to a Gaelic school. Two very friendly ladies, speaking a confusing blend of English and Gaelic (confusing to us, they seem to be getting on just fine) welcomed us there and served excellent soup, tea and home bakes, including the allegedly 'traditional' Bakewell tart—I wasn't at all sure it was a Hebridean tradition but they seemed excessively keen that we should try it and I seldom like to disappoint where cake is concerned. It was indeed an estimable comestible.

The museum is in an adjoining building, beside the tea room. It's operated by the Comann Eachdraidh Uig (Uig Historical Society) and seems to consist mainly of static displays of local archaeology, most notably some replicas of the famous Lewis Chessmen, or Uig Chessmen as they are proudly styled here, Viking life, crofting and so on, all fronted by a curiously life-like mannequin of a lady reading a book. Closer examination, however, revealed that she was no mannequin but a volunteer, who appeared to be asleep.

It seemed churlish to wake her and as my wife has a near pathological distaste for museums on account of having been dragged round any number of them during an 'improving' childhood, we stepped outside to inspect Stephen Hayward's fine wooden figure of a Uig Chessman which stands guard at the entrance. A partner piece may be found at the original find spot on the machair at Ardroil, though some sources maintain the original spot to be at Mealista—perhaps this is why the dispossessed and uprooted chessmen still look so very glum.

Our arrival at Abhainn Dearg was itself certainly glum. It was soon apparent that, despite the visit having been arranged by email, or so I thought, we were not expected and the distillery gates sported a large 'CLOSED' sign. Ignoring this, I drove confidently in, and parked opposite some drab and unprepossessing buildings where we were met by Laura, who seemed less than keen to have her name included in the book; doubtless she has her own reasons. While perfectly amicable and more than happy to help, she was understandably nonplussed at our presumption in ignoring the sign. We chatted for a while, completely at cross purposes.

After some further confusion and an increasingly desultory conversation punctuated with anxious glances towards the office, she went to fetch the distillery's founder and owner, Mark Tayburn, who in a further twist was on the phone and evidently not expecting us, though Laura did a fine job in concealing that as we proceeded to the maltings.

And this is where the fun starts. You must understand that Abhainn Dearg is not like any distillery you've ever visited or will ever visit, unless you spend much time in the company of the criminal classes. Whilst Abhainn Dearg is legal, fully licensed and entirely above board it is also the closest you will ever come, or probably want to come, to an illicit distilling operation. Imagine an abandoned fish farm with all the glamour of a failing Siberian tractor collective. Hold that picture.

For one thing, it has the authentic island air of disarray and abandonment that I recalled from my childhood visits and which, I will admit, gave me considerable pleasure to see. Bits of equipment, some evidently relevant and others such as a broken outboard engine not immediately so, were scattered at random. Everything looked homemade and temporary. There was a noticeable lack of health and safety notices and the regulation government signs that bedeck other sites.

No one appeared to be wearing standard issue company clothing. Our friendly guide, now on familiar territory and more animated, was in jeans and a smart blue sweatshirt decorated with the Uig Chessmen, who weren't in the slightest looking any happier. But I thought it was all perfectly agreeable and cheered up enormously, all the initial awkwardness behind us.

The malting consisted of a small room containing a large open tray or trough on legs, under which there was an old wood-burning stove (aka an exothermic oxidizing reactor). This was, curiously enough, burning wood as we stepped inside, the heat from which was directed to the tray on which was spread a quantity of freshly malted barley. Faint wisps of steam rose upwards.

A day before, this barley had lain on the floor slowly germinating, just as it should, roughly where Laura was now standing. So, despite the apparently ramshackle arrangements, this was a perfectly functional malting, albeit one totally reliant on manual labour at every stage. As, indeed, every malting would once have been.

What I couldn't work out was how the malt was peated, but the explanation was simple: apparently, the fire was loaded with peat, the flue disconnected and the door closed, at which point the room filled entirely with smoke, some of which was absorbed into the malt. It was, I was forced to concede, an effective if rudimentary system, though singularly unpleasant for the first person to return to the smoke-filled room when more fuel was needed for the fire or the process was thought to have gone on long enough.

Of course, peating levels will vary from batch to batch, but if variability is something you prize then this is hardly a problem. What was both impressive and gratifying to know is that the barley itself came from Mark Tayburn's own fields near Stornoway, making this endeavour truly local. He is currently growing and malting Golden Promise and Concerto varieties from his own fields. I hope and expect even older heritage strains, perhaps some bere, will be trialled in the future.

At around this point, explaining disarmingly that he 'never reads emails' a profusely apologetic Mark materialised to continue the tour and Laura moved happily on to look after a confused-looking group of English tourists. Presumably they were expecting a corporate welcome and big production set-up; the homespun and very Lewisian arrangements seemed to unsettle them and they departed with undue haste. It's a shame; they will never know what they missed.

Onward then to mashing, fermentation and distillation, all housed under one roof in an adjacent building, also more functional than decorative in architectural style. The stark and uncompromising façades of these facilities will come as a shock to anyone accustomed solely to the more sanitised, picture-postcard surroundings of Scotland's larger distilleries—which is to say, the vast majority of them.

But this must surely represent the purest, most primitive form of legal distilling that's it's possible to see anywhere in Scotland, and thus the single step that I took from the courtyard to the still house not only vindicated every one of the sixty awkward and tiring miles of the drive but acted as a kind of time machine to a simpler age.

If the replica Sma' Still at The Glenlivet, which on Speyside Festival days is fired up for the curious, is a gateway to the smuggling of illicit drams across the Cabrach, then the Abhainn Dearg stills represent the point of transition from a furtive and clandestine craft tradition of skulking in bothies in remote glens beyond the reach of authority to the early first mechanisation and the beginning of mass production. Think of this as 1823 if you will, and try to imagine a distiller acquiring his first licence and taking the hesitant step to legitimacy and the public acceptance of more than a strictly parochial and local community. Abhainn Dearg is therefore a distillery of an importance and significance considerably greater than its modest scale and, I would submit, of more consequence than it actually realises. This is a distillery still expressing the original soul of whisky and the robust, self-reliant spirit of this island. It is a rare and precious thing.

Not that everything here is make do and mend. The stills may bear a curious resemblance to repurposed hot-water cylinders but Mark proudly led us to one of the distillery's rambling sheds which houses a discreetly located but sophisticated, modern hydro-electric power scheme. Water is 'borrowed' from a hundred yards or so up the Red River, passed through a turbine, electricity generated and the water returned to the main stream.

And here two arms of government come into conflict. The environmental agencies are uncomfortable with even this modest level of water extraction while the enthusiasts for green power (and the Scottish Government has been a righteous and earnest advocate of such benign sources of energy) are urging Mark to do more with his scheme. Wanting to move forward, he is trapped between two conflicting agencies of the state. As he points out, fish will only be running up the river when it is in spate and thus higher levels of extraction may easily be sustained.

Mark also proudly noted that his small group of buildings was the first and only human construction on the river. Behind the site lie the mountains of Brinneabhal and Cleite Leathern and the Red River, fed by storm-driven Atlantic clouds that have not seen land for more than three thousand miles, runs through open moorland. There is no habitation and no agricultural activity on this land; arguably, therefore, this is as pure a water source as may be found anywhere in Scotland.

There is a modest tasting room here, where you may sample and buy the Spirit of Lewis and the Abhainn Dearg single malt.

The new-make spirit is heady stuff. I can imagine it taken from a tin cup in some remote bothy after a hard day in the mountains, or to provide the energy to pull from the kelp-draped-deep creels laden with sea-blue lobsters. Aeneas McDonald's immortal words came to mind: 'there are flavours in it, insinuating and remote, from mountain torrents and the scanty soil on moorland rocks and slanting, rare sun-shafts'.

Moorland rocks and slanting, rare sun-shafts speak eloquently of Lewis. The landscape is strange and unfamiliar, with bare rock poking everywhere through the thin soil. This is a harsh and unforgiving land, not without its own stark and austere beauty, but composed of the same granite that makes up the moon.

Mark left us abruptly. A friend's bull had broken out of its field and he was needed as part of the rescue party. It was the perfect end to a visit that was itself perfect in so many entirely unexpected ways.

Back in Harris we paused at Bunavoneader to visit the old Norwegian whaling station, abandoned in the 1950s.

Today it has been roped off in an attempt to prevent damage to the fragile remains, with a polite but firm notice requesting visitors not enter the site. That's a perfectly reasonable appeal, which we respected, but I was sorry not to see once again the headstone to Sam, the 'faithful dog' of Captain F. Herlofsen, which I recalled from earlier visits. The Herlofsen family owned and ran the station and, in the autumn of 1907, Captain Herlofsen was presumably so distressed by the death of his dog Sam that he saw fit to erect a fine memorial, which today has endured longer than much of the whaling works.

In its late flourishing, the station at Bunavoneader processed some thirty whales annually, but still proved uneconomic. In its turn-of-the-twentieth-century operations, however, it was said to provide the highest yield of whale oil from a single station outside of Iceland and at its height employed nearly 160 locals in processing the catch.

Over the years, Fin, Sei, Right, Sperm and even Blue Whales were landed here. The largest whale landed was a Blue, brought ashore in the 1950s and, presumably amongst scenes of some jubilation, rendered down to whale meat and assorted by-products. Perhaps even a celebratory dram or two passed the workforce's lips, or possibly, even more poignantly, it was simply greeted with a sigh of resignation at the additional hard work it represented. The contrast between the abandoned ruins, not to mention the whales who died to feed the rendering plant, and the headstone to Sam the faithful dog needs no commentary.

Today this is a scheduled ancient monument which the North Harris Trust are working to preserve. Eventually it will re-open to visitors and the casual and curious can learn something of this industry, once seen as so necessary and vital and which today would be thought cruel and barbaric. Along with the Hebridean basking shark fishery, which also flourished for so long at such a cost, whaling belongs to our past, but is not a part of our history to be quietly forgotten.

Today's visitors, already some 68,000 of them annually, flock to the Isle of Harris distillery at Tarbert. There could not be a more vivid contrast, however, with Abhainn Dearg. Harris is the very model of modern, marketing-led distilling.

The project, styling itself 'the social distillery; for, with and from Harris' was initially the vision of New York musicologist Anderson Bakewell, sometime resident of Oxfordshire but long-time owner of the island of Scarp, one of the small islands that ring Harris and Lewis. His goal was to 'create a catalyst for growth in the Hebrides', an ambitious aim that has brought others to grief in the past; one thinks, for example, of the noble Lord Leverhulme before him, who had attempted without success to revive the

Bunavoneader operation with a bizarre but unsurprisingly short-lived plan to sell tinned whale meat sausages to Africa.

So far, though, things are going well. Albeit heavily reliant on public sector funding, the £11 million project is up and running and today is producing whisky and its unexpected success, Harris Gin. Around twenty people are employed and there has clearly been a beneficial impact on awareness of Harris and to locally based tourism businesses.

Everything here is shiny and new, every 'touch point' reflects the corporate ethos, the earnest statements of values and the brand positioning model that so clearly underpins this enterprise. From the design of the very shelves to the stock they carry, everything has been painstakingly curated to reinforce a pre-determined image. Even the menu in the café (excellent, it must be stressed) is resolutely on message but, after visiting Abhainn Dearg, it did feel just a little over-controlled, managed and sanitised. This, I concluded, is how Walt Disney would build a distillery.

But, like Disneyland, the management very clearly knows exactly what they are doing and why. For the typical tourist visitor, this will be delightful and the product will be good, consistent and palatable, whether that product is the liquid produced here, a quick cappuccino while waiting for the ferry or the overall 'visitor experience'.

That asks little of the visitor other than the willing suspension of disbelief. It would be cynical of me to question whether close to £5 million of public funds, albeit in different guises, is appropriate for an industry which has excess capacity or is an efficient way to create twenty jobs; it would be churlish to observe that of the seventeen private sector investors few are resident on Harris and thus to question the claim that this is 'for, with and from Harris', for, if the eventual profits flow off the island, what enduring benefit is there other than some new pay cheques, welcome though these undoubtedly are. It is perhaps unfair or unreasonable to cavil, for so much has clearly been achieved.

For the most part, the senior management is not for, with and from Harris, though to be fair again, the skills to secure investment of this scale, write the plan, design, install and operate the plant and market the product were never going to be found here, and outside expertise was clearly essential if this was ever going to work. But where Abhainn Dearg feels as if it has sprung from the earth, the Isle of Harris distillery seems to sit heavy upon it, however elegantly.

What of the product though? The stills at the heart of the distillery are Italian. No Scottish manufacturer was able to supply in the time required it seems, something that is regrettable but wholly credible given the current pressure of demand for new stills from their large and established customers. So, to Frilli of Monteriggioni in Tuscany went the order and this one-hundred-year-old, family-owned company built their first stills for the Scotch whisky industry. We shall gloss over the fact they were paid for, in part at least, by Scottish taxpayers, for we are, after all, all good Europeans now — as First Minister Sturgeon is ever eager to emphasise.

An important part of the design of the stills lies in the neck and lyne arm, where a purifier has been introduced to potentially increase reflux to the still, thus giving the distiller greater flexibility in selecting the final spirit character. Whether the idea for this came from Ardbeg I cannot say, but the senior management at Harris had previous experience of that distillery so it is not inconceivable. Eventually it is planned to produce lighter and heavier versions of the spirit, and a heavily peated variant will also be introduced — with only one pair of stills to work with, such versatility will be essential if a range of expressions are to be offered.

Where the distillery has had an early and unexpected success has been with its Harris gin, a beneficiary of the boom in craft distilling and the largely unexpected revival in gin sales to a new, younger consumer. Like Bruichladdich on Islay with their offering, The Botanist, Harris is riding this entirely fortuitous wave of good fortune with all the considerable sales and marketing skills seen in the rest of their operation.

At the time of this visit around 40,000 bottles had been sold, which, from a standing start and with only some public relations work and a deft social media campaign to create awareness, is actually an impressive total. The bottle is, I have to say, extremely handsome; a pleasure both to behold and to actually hold. It is not every new start-up that can afford top designers and a bespoke bottle mould, and not every top designers' bespoke bottle works in the market, but Harris have spent their money shrewdly and are reaping the rewards; justifiably so, as the contents do live up to the promise of the packaging.

There is, I suppose, a danger that if the gin continues to sell well and profitably (and at £35 a bottle with a shrewd direct sales business model it is certainly very profitable), that Harris will become better known for its gin than its whisky. However, this could safely be categorised as not the kind of problem that will unduly trouble the happy investors, for all the protestation that this is at heart first and foremost a whisky distillery. What's wrong with gin, anyway?

All new gins make great play of their unique botanicals and Harris is no exception. In this case the unique ingredient is hand-dived sugar kelp, or Saccharina latissima, to give it its Sabbath name. This was the idea of the distillery's consultant ethno-botanis, Susanne Masters, who was exploring the role of local botanicals. Included on the basis that collecting the crop by hand using a diver was sustainable, the seaweed adds a note of salt but, unexpectedly, some sweetness to the final spirit.

Diving for seaweed. It's a living, I suppose, but probably not what any careers master would recommend. Incidentally, once upon a time, kelp was a valuable Hebridean crop and substantial quantities were harvested and shipped to Greenock and Liverpool, where they were burnt for the soda ash required in making glass and soap. For example, in a little over six months in 1792, more than 1,800 tons of kelp was recorded as leaving Tobermory alone. However, that industry collapsed in the 1840s when synthetic methods of production took over.

Today, valuable alginates are extracted from kelp and I've also seen it suggested that large open-ocean kelp farms could serve as a source of renewable energy. Perhaps kelping wouldn't be such a poor long-term career choice after all.

If you can't detect the sugar kelp influence in your gin and tonic with absolute certainty (and why should you—it's hardly a familiar flavour note) the distillery make it easy for you by offering a concentrated extract, the imaginatively named Sugar Kelp Aromatic Water (£20 for a 50ml bottle and dropper. I've paid more for ink, though not often, and I don't add that to my gin). You add this to your glass for added sugar kelp flavour.

This elixir has been prepared for them by Amanda Saurin, a self-styled Apothecary, in what she describes as 'a slow, meditative process, a precise art that has been passed from hand-to-hand, drawing on the alchemy of intuition and the wisdom of generations of healers, travellers and makers'.

That seems, to my worldly and insensitive soul, uncomfortably similar to the ancient trade of selling snake oil to the gullible with artfully crafted jibber–jabber and New Age incantations. The 'alchemy of intuition' indeed. Sounds great until you ask what it might mean.

Does it change the taste of the gin? Well, yes, it certainly does, and 50ml dispensed by dropper goes a long way. Using the dropper (bullshit alert: marketing jargon coming up) adds to the ritual of the serve and if there's anything that a right-on twenty-first-century marketing person loves, it's the ritual of the serve, particularly with added alchemy of intuition.

Does it make the gin taste any better? Well, yes, it does if you think it does, and that's probably enough.

The Isle of Harris distillery, very, very well done as it is, represents the apotheosis of island as lifestyle accessory: the new designer-cool version of sanitised wilderness packaged (to paraphrase Amanda Saurin) to capture the imagination and feed the soul. This is Harris seen through a Boden lens. The mantra, it seems, is that on Harris there is 'time to be'.

Always assuming you could get time off from the whale killing.

I saw a harder-edged Harris in the 1960s. It was a favoured destination for family holidays, time largely spent fishing and, as I recall, eating sandwiches which fully lived up to their name. There was probably a tartan Thermos flask if I think hard enough about it. There certainly should have been.

My mother, from solid Potteries blue-collar stock, never seemed to me entirely at ease here, though insisting loyally that she was enjoying herself. She maintained that shopkeepers raised their prices for visitors and, bred in the tradition of Midlands England working-class solidarity and hospitality, was greatly offended when locals switched into Gaelic as she would enter the local stores.

'They were speaking in English until they saw me,' she would insist, 'but then only in Gaelic.' For some years afterwards she held on to the idea that she would learn a few key phrases in Gaelic and then deploy them loudly on leaving the shop, spreading confusion in her wake.

'Tapadh leibh,' she imagined herself saying. 'Tapadh leibh. A tlachdmhor latha a tha thu a h-uile', which Google Translate assures me means, 'Thank you. A pleasant day to you all.' (Honestly, if I didn't trust Google it could be Klingon for all I know.) But possessed of a bottom of common sense she knew she would never carry it off, or risk being challenged and found a pretender, so the plan remained solely to comfort her imagination.

My father had greater success in cultivating some locals, perhaps because we were paying good money to rent the damp and cramped cottage that they had sensibly moved out of, or more probably because he sweetened the relationship with quarter bottles of whisky bought from the self-same store, where, curiously, he seemed to have little difficulty in making himself understood.

The whisky was to induce friendly locals to point out favoured fishing spots and we soon became particular favourites of one worthy. In my memory, he is named Donald Macleod, which is credible enough, though not to be relied upon in court proceedings.

Several quarter bottles had changed hands and we had enjoyed productive sport in a sheltered inlet near Rodel when Donald abruptly volunteered to show us how folks really fished on Harris, an offer accompanied with a manly wink to father, who little realised at that stage what he had let us in for.

We were inordinately pleased with our Abu Toby lures, then pretty much state of the art in sea-fishing spinning tackle, so were more than a little put out and confused to be told that our 'fancy rods and spinners' could be left in the car. But, thoroughly mystified, duly excited and a little apprehensive, we met in the appointed spot and set off in an alarmingly small and decrepit rowing boat. Some time was spent in ensuring we were alone and out of any immediate line of sight.

Donald pulled strongly enough on the oars but took us only a few yards offshore to a point between the shore and a small group of rocks. Here we stopped, drifting only slightly in the slack water while our guide fumbled in a box under his seat. From this he removed an army-issue grenade, presumably left over from

Second World War exercises, removed the pin and with studied casualness and an economical arm lobbed the ordnance casually into the sea.

A few seconds later the result was predictable: stupefied, effete visitors from Edinburgh watching slack-jawed as dead and stunned fish floated to the surface of the water. The larger of these were swept up and thrown into the bottom of the boat (no place for flowing skirts, as you will learn). 'Dinnae bother with tiddlers, lads,' we were commanded—and there was no need, for fishing in the Harris style had proved productive. We were now, of course, in flagrant breach of the Fisheries (Dynamite) Act, 1877, a fact of which Donald seemed curiously indifferent, though I understand that islanders talk of little else during the long winter nights.

The remainder of the evening was spent in driving to various cottages and small crofts while Donald handed out free fish. As I recall, no questions were asked, until we returned to my mother, who seemed to have more than a few for father, who, in turn, seemed more concerned about the state of his car boot.

In the interest of full disclosure, I must add that, now aged ninety but mentally as acute as ever, father claims to have no recollection of this remarkable event. Either it has been selectively repressed by him as a commentary on his parenting or I am suffering from false memory syndrome. Whatever the explanation, I fully expect to hear from some authority or other. I propose to explain to them that the fish are all dead, and in all probability so is Donald, if that was even his name. But I remember the evening very clearly and having shared it recently with several islanders, native Hearachs all, who found my tale entirely credible, have no doubt that it happened and that father has suppressed it for very sound and wholly understandable reasons. Any responsible parent would do the same. I can hardly imagine the interest that various caring and concerned agencies of government would take today in such an adventure, not to mention the response of their colleagues in the Anti-Terrorism squads.

That Harris of memory seems a foreign country. Rodel is still there, of course, with its famous church and the Macleod tombs. Being unable to locate the source of my piscatorial adventures, we visited St Clement's and marvelled at the ornate carvings, and the freedom with which we were able to access the church and these remarkable monuments, exactly as we had earlier walked freely up to and through the ancient standing stones at Callanish.

But as we drove around the island I continually felt that something was missing. Finally, I arrived at the answer: gone are the abandoned cars and rusting machinery of youthful memory that seemed to litter every field and derelict steading. Nothing, it seemed to me back then, was ever cleared away, but all was left to rot and decay, lending the island an air of sorry neglect. Islay, as I recall, was much the same.

Since then, across the Hebrides, someone, whether local council or some enterprising mogul of the scrap metal trade, has done a fine job in clearing all this. There is scarcely a rusting old vehicle to be seen and those few which remain come as something of a shock. This care for the local environment, whether driven by social enterprise or the profit motive, is a thoroughly good thing.

What did both shock and amaze me was the preponderance of luxury self-catering accommodation to be found on both Harris and Lewis, manifested in remarkable *Grand Design* properties such as the astonishing Beach Bay Cottage, located a short distance from the Abhainn Dearg distillery.

Essentially a high-tech tribute to the earliest Neolithic designs or the black houses of Hebridean tradition but with every creature comfort imaginable (black houses rarely featured underfloor central heating or a private sauna), this exceptional property rents for up to £2,500 per week in peak season and sleeps just four lucky guests. It is not the only example, but the juxtaposition of this luxury home against its near neighbour, the starkly functional Abhainn Dearg distillery, brought home very vividly to me how the island had changed and was continuing to change. When last I looked, Beach Bay Cottage was all but fully booked up to a year in advance.

How, then, shall we leave Harris and Lewis? The contrasts here between the islands I remember and those that I found in 2016 were profound, and the contrast between Abhainn Dearg and the Isle of Harris distilleries was both deep and, I sensed, rooted in a set of almost incomprehensible and probably unbroachable cultural differences.

But perhaps Abhainn Dearg will move more to the mainstream and become more corporate in its approach. I wonder at the tolerance shown to it by Her Majesty's Revenue and Customs; not, in general, a body noted for their flexibility or acceptance of an independent and buccaneering approach to life. Few distillery managers would leave us, as Mark suddenly had to, to help a friend recover a bull which had broken out of its pen, but then few distilleries are as close to their agrarian roots as this one.

Perhaps as it matures, the Isle of Harris distillery will weather and grow to be more like its eponymous home; perhaps something of the island will pass into its soul as the new-make spirit slowly turns into whisky. At the moment, for all its protestations of local identity, it seems to me to sit as an outsider, transplanted from some alien corporate culture, wearing its smart new designer clothing rather awkwardly against the background of Harris Tweed.

I struggled for some time as to how to convey this in my manuscript. Eventually though, the stark, harsh granite outcrops with their sharp peaks were well represented by the narrow nib of a Pilot Vortex, a relatively modest Japanese pen which made a strangely satisfying cutting noise as it transferred my words to the paper.

The ink was a problem until I discovered an unusual French product from a Paris-based manufacturer of sealing waxes and inks, J. Herbin. Their 1670 range commemorates the founding of the firm with three distinctive coloured inks, each containing very fine metallic particles, which glint as the ink dries.

I chose to work in their Stormy Grey, the austere tone of the base anthracite ink echoing the granite I saw around the island, but the metallic speckling catching the light just as elements in the stone caused sudden and unexpected glints from those slanting, rare sun-shafts.

That may seem an extravagant and fanciful notion, but this ink has a personality that reflects the quality of Harris and Lewis and the ancient stone of which they are formed. Ink as metaphor—it's an intriguing point on which to step back onto the ferry and return to tourist-crammed Skye, home to the mighty Talisker and its companion-to-be at Torabhaig, via a short diversion to the little island of Raasay, which was somewhere entirely new to me.

9

Raasay

Nothing to see here, says the Great Cham

IF YOU DRIVE PART OF THE WAY ACROSS SKYE on the main
A87 towards Portree you will pass signs to the Sconser ferry, from
where it's possible to go over the sea to Raasay, a small island just
off the coast of Skye.

Unfortunately, according to the Great Cham (Dr Samuel
Johnson himself), 'Raasay has little that can detain the traveller,
except the Laird and his family'. However, those travel notes dates
from 1773 so we decided we could disregard them and go anyway.

For the magnificent sum of £3.70, a foot passenger can make
the return trip as we did, though not necessarily accompanied
by a group of excited school children, en route to a field studies
day on the little island. How exciting to escape humdrum,
everyday Skye for the romance of Raasay—or so I presume it

must feel if Skye represents the limit of your everyday existence.

Raasay has not hitherto troubled the distillery historian. It is not recorded in H. Charles Craig's exhaustive *Scotch Whisky Industry Record*; Misako Udo's *The Scottish Whisky Distilleries* omits any mention; and though Alfred Barnard must surely have passed very close to Sconser on his visit to Skye, and his steamer stopped briefly on Raasay, he was travelling with a firm determination to make his report on Talisker and, as he tells us, was anxious about the weather (which turned out nice as it happens). Accordingly, he neglected to disembark or later make the short crossing to Raasay. In fact, he may not even have given it more than a sideways glance, as his gaze was directed to Skye's more imposing vistas. He was like that, forever looking at the unexampled grandeur, the romantic and wild mountain torrents and the sublime prospects of the desolation of a silent wilderness, after which he would frequently quote some apposite verses. Scott was a great favourite.

His unaccustomed reticence, of course, was for the very good reason that there was then no distillery on Raasay; or no legal one at any rate. However, like many a Hebridean island, there is a tradition of illicit distillation, which in this case has been recorded through a study of Gaelic place names.

In her book *Gach Cuil is Ceal* (*Every Nook and Cranny*), Rebecca S. Mackay discerns the dim history of illegal distilling on Raasay and the neighbouring smaller island of Rona that lasted until around 1850. It is written in place names such as Taigh na Poiteadh Dhuibh (Home of the Black Still). That seems quite explicit and Mackay also suggested that distilling was a frequent occurrence at Screapadal where Vamha an Ochd Mhoir (the Cave of the Big Hollows) was a spot particularly favoured by topography, giving the moonshiners a clear view of any approaching excise officers. It is said too, that friendly crofters on Skye would signal to Raasay should any government men be seen drawing uncomfortably close to the island.

Detection would have been a real problem for the bootleggers for, as Mackay relates, that would have provided a hard-hearted laird with all the excuse needed to evict the miscreant tenant and put sheep in their place, not that much excuse seems to have been required. But the distillation of any spare barley or, more correctly, bere, would have been a time-honoured tradition. Some of the *uisge beatha*, taken straight from the worm (the coiled tube that acted to condense the spirit), may have been used to pay rents but, in general, the small production was for household use only—a necessary prop against the damp and cold of a black house.

On Skye, the late Sir Iain Noble honoured this tradition with his company, The Gaelic Whiskies (Praban na Linne Ltd), and would volubly maintain when asked, and even when not, that he 'would neither confirm nor deny' that his Poit Dhubh malt came from an illicit still. Like so much that flowed so freely from the irrepressible Sir Iain it was all the most frightful nonsense; not least because the Poit Dhubh was blended and bottled by the completely respectable Broxburn Bottlers, who would never have permitted a drop of contraband spirit to cross their door for fear of losing their licence and thus their whole business. Not to mention the ill-concealed schadenfreude of the rest of the distilling industry.

Sir Iain who, in my experience at least, was a wily and challenging fellow with whom to do business, died before he could see his dream of a traditional farmhouse distillery at the Torabhaig steading, on Skye's Sleat peninsula. But it has finally been developed in a form closely resembling his remarkable vision and it was my great pleasure to visit it later on this trip.

The Torabhaig project probably more closely represents—albeit on a larger scale—the type of artisanal distilling that would have taken place on Raasay. There was certainly never any industrial distilling here and, as Mackay surmises, the farm tradition probably died out in the mid-nineteenth century. But though he respects his Raasay forebears, Alasdair Day is aiming altogether higher and it was his nascent Raasay distillery that we had come to see.

Alasdair is most personable company, yet quite seriously driven. I met him first when he worked for a dairy company and was involved in large-scale cheese production. His background is as a food scientist but his family were, on his great-grandfather's side, licensed grocers. This is, in Scotland at least, an honourable profession, once charmingly but confusingly known as 'Italian warehousemen'. John Walker was a licensed grocer, as was the first Dewar and so too the Chivas brothers, all names on well-known blended whiskies to this day.

Day's ancestors were not as distinguished, nor presumably as successful, but his great-grandfather, Richard Day, had a small licensed grocer's shop in Coldstream in the Scottish Borders where, like many other similar operations, he blended and bottled small quantities of rum and whisky. In particular, he created something he called the Tweeddale Blend, recording the precise recipe in his private journal.

That business eventually closed and the records were all that remained, passed down through the family as something of a curiosity. Like many a late nineteenth or early twentieth-century blend, The Tweeddale could well have been coloured, sweetened or fortified with the addition of a little sherry, brandy or even rum, but when in 2009 Alasdair Day decided to recreate the blend it was done strictly in line with the current regulations. They, of course, absolutely prohibit any additions and, as a trained food scientist, Day naturally eschewed the ancestral legerdemain of such furtive adulteration.

I encountered him shortly after he launched his recreated Tweeddale Blend as a tribute to the family heritage. I thought it a perfectly pleasant drop of whisky but more of a curiosity that seemed an improbable basis for a long-term, viable business. After all, a premium blend without heavy advertising support, no brand heritage to speak of and lacking the distributor muscle of the major operators could surely be no more than a one-off novelty and, after the initial fun and excitement wore off, Alasdair would surely return to corporate life and the high-octane world of cheese making.

Well, I evidently failed to gauge the man with any accuracy, for it seems his love for whisky exceeded his ardour as a turophile, and it was he that I boarded the ferry at Sconser to meet, prepared to see the building works on his Raasay distillery.

And they were, I have to admit, impressive. Some serious money is involved here. Alasdair was coy as to the exact amount, or even the vague amount, but it's clearly a significant investment. This hasn't been achieved via crowdfunding, or the speculative sale of casks that might never be distilled, but through hard-nosed investment, much of it from his business partner, Edinburgh-based Bill Dobbie, an experienced entrepreneur who has provided a significant portion of the initial capital expenditure.

We arrived to the sound of machinery: bulldozers and some terrifyingly loud and very large earth-moving equipment was in frenzied action. It was clear that much had been done to clear the site, which sits a few hundred yards above the little ferry terminal. Previously, it had been a typically Victorian mansion house and subsequently a hotel, which, like so many others, had proved uneconomic. The twin pressures of a relatively short season and high cost of maintenance eventually drove it to the wall. With visitors expecting ever-higher standards of comfort and more sophisticated facilities, it was doomed.

Earlier owners had tried to expand in an effort to achieve a viable scale, adding the usual hideously inappropriate extensions that characterised much 1970s development, to the point where the original Victorian villa was all but obscured. I suppose all one can say in defence of the additions is that at least they didn't have flat roofs. If I envy Alasdair anything in this project (and I certainly don't envy him the inevitable stress and sleepless nights), it would be the pleasure of removing these ghastly additions and revealing the Victorian heart of the building. I would happily swing the wrecking ball myself.

However, with the site clear, the potential was immediately apparent. The original home will serve as a clubhouse for members of the Na Tusairean Club—this is an opportunity to purchase one of the first 100 casks and enjoy on-site accommodation in some luxurious rooms that will be provided for members. If you can realistically see yourself visiting Raasay on a regular basis it's clearly an attractive proposition.

To one side of the house, with its elegant members' tasting room, will be a visitor centre, as naturally tourism is at the centre of this project, and to the other side is the distillery itself, in a strikingly modern construction. On the slope immediately above all this will be a warehouse and modest bottling complex.

Being entirely honest, I don't think that I care overmuch for the visual impact of the distillery. It's an unashamedly stark and modern building, with no attempt made to soften its overall impact or to copy traditional building styles. However, this was what client, architect and, above all, the planning authority wanted and it does have the very great merit of not apologising for itself or appearing as a pallid pastiche of the vernacular with some faux pagodas pointing pointlessly at the sky. For all that, I have no idea how it will age...[1]

As I write, the stills and all the associated pipe runs and other vessels are being installed. Like the installation at the Isle of

---

[1] This, as I'm sure you appreciate, is an example of the rhetorical artifice known as aposiopesis. Impressed? Sadly, I don't suppose that you are.

Harris, most of the equipment has come from the Italian firm Frilli who, though they have manufactured stills since 1912, must be thoroughly delighted with the explosion of new distilleries in Scotland. Raasay Distillery has an excellent website with time-lapse photography demonstrating progress on the construction and they are active on social media, to allow anyone interested to keep up with their impressive progress.

The distillery will be working by the time you read this, as they hope for an official opening in September 2017, with distillation having started.

It is not an enormous plant, though. Initially the production is estimated at some 100,000 litres of pure alcohol annually. That's typical of these boutique distilleries but, with 100% of the output said to be reserved for the company's own use (apart, that is, from those members' casks) there will still be a considerable quantity of stock to consider in three to five years' time, let alone ten or twenty years hence. Right now, they don't actually have a distiller in place, but have advertised on social media for applications, stressing the unique nature of life on Raasay. Apparently, applications have come from far and wide, including some from Canada and parts of Africa, which sounds like an interesting cultural challenge for someone.

But if all goes well and they do appoint a distiller, from some time in 2020 they aim to sell around half the production, or some 150,000 bottles each year, reserving the balance of the stock for long-term maturation. That's quite an ambitious target but sales and brand awareness are being created by clever social media campaigns as with the recruitment of a distiller and by the release of teaser expressions such as Raasay While We Wait, which is a single malt from another distillery, obviously, that aims to showcase the desired style.

A branch office has also been set up in Taiwan, which has a

significant and buoyant whisky market with some very well-informed consumers as well as two single malt distilleries of its own. In addition, supplies have been shipped to China and the Philippines, so the global ambitions of this fledgling venture have been set on solid foundations. Having tasted the target style, I've every confidence that Raasay single malt will be a useful addition to Scotland's distilleries and will find a place in drinker's repertoires.

From the Tweeddale Blend to Raasay Distillery, Alasdair Day has made unusually swift progress in the whisky business, though he's modest enough to regard what he's achieved with some humility. But it's actually part of a larger plan, as Raasay is only the first, and smaller, of two distilleries that the company intend to build. The second will be located in Peebles, in the Scottish Borders; hence the company's name R&B Distillers (it stands for Raasay & Borders).

I was actually surprised that they were able to register that identity as for many years R&B was the name by which everyone in the whisky trade knew the influential Glasgow blenders Robertson & Baxter, by whom I was once briefly employed. However, today R&B has been largely subsumed into the Edrington Group and awareness of a once potent, highly regarded and well-established member of whisky's aristocracy has gradually faded. So perhaps no one minds, but R&B Distillers carry a proud legacy, even if that was not their original intention.

I fully expect the Raasay project to be a critical and commercial success. The people behind it have a solid track record of achievement in business, the project is apparently well funded and the site — clearly visible from the approaching ferry and a short walk from the harbour — will be a tourist magnet. At present, Skye cannot easily accommodate the visitors it receives during the peak season. Talisker gets uncomfortably full and the Torabhaig distillery is not really designed with large number of visitors in mind. The regular visitor to Skye may well welcome the chance to step off the island, at least for a few hours, and whisky enthusiasts—distillery baggers to a man—will doubtless be keen to add Raasay to their list of distilleries visited and whiskies tasted, carrying away a bottle or more in triumph.

Sadly, time did not permit any further exploration of Raasay. We left, determined to return and impressed with what we had seen. If I felt confident of regular visits I would seriously consider joining the distiller's club and making this my home from home, or at least an agreeable base for further exploration of this charming little island.

I would seek out the rare and elusive Raasay vole; climb Dun Caan for a view of 'my' distillery; visit the deserted settlement of Hallaig and the ruins of Brochel Castle (Daniell's evocative engraving is a powerful lure); and explore the unique geology or follow in our royal family's footsteps with a picnic on Inver Beach. Apparently, the Royal Yacht *Britannia* was known to anchor here and the Queen is rumoured to have enjoyed the seclusion provided by the beach, far from the long lenses of the prying paparazzi.

Incidentally, our present royals are not Raasay's first aristocratic visitors. Though such is the power of song that we hear much of Prince Charles Edward Stuart visiting Skye, I was fascinated to learn that he also landed on Raasay on 1st July 1746 as he escaped from the Hanoverian troops seeking his capture.

It's probably just as well that he didn't linger, for it must have been a depressing sight. Around one hundred of the MacLeods of Raasay, along with twenty-six pipers, were out with Charlie. Though all but fourteen returned home the island did not escape the vengeance of the Duke of Cumberland's troops.

Bishop Robert Forbes, a Jacobite, was told that 'the whole island of Raasay had been plundered and pillaged with the utmost degree of severity, every house and hut being levelled to the ground, and there was not left in the whole island a four footed beast, a hen or a chicken'. That's a war crime, by the way, and would be thought very bad form today.

But within a few years the MacLeods had rebuilt Raasay House and were able to entertain Boswell and Johnson on their Hebridean travels. They visited in happier and more prosperous times. Though Johnson seemed to feel there was nothing to see here apart from the laird and his family, he was generous enough to elegantly liken the hospitality he received to that extended to Ulysses in the Odyssey when he is shipwrecked on the island of Phæacia and meets the princess Nausicaä, surely the most gracious of compliments, though one unlikely to be found in today's Rough Guide or on TripAdvisor.

A few years later, Raasay found wider fame in the illustrations produced by William Daniell, who visited Skye and Raasay in July and August 1815, including a number of views in his best-known work, A Voyage Round Great Britain. His work is a reminder that the best sense of many of these islands may be gained from the sea, though at the same time these images contributed to the widespread and long-standing image of the Hebrides as a romantic playground.

And in all fairness, I should conclude that the Great Cham, Sage of Lichfield, was mistaken. There is much, as I learned in my all-too-brief excursion, to detain the traveller and I hope one day to return and report on the distillery at work.

However, after this scantiest of visits we rejoined the chattering schoolchildren and flitted gently back over the sea to Skye and on to Talisker and Torabhaig.

10

Skye

Shopping with tokens

IMAGINE, IF YOU WILL, that you are the energetic and entrepreneurial agent of a major landowner—let's call him John MacLeod of MacLeod—on the island of Skye. We're in the reign of King George IV and, following the end of the Napoleonic Wars, things have begun to go badly for the Hebridean kelp industry. Landlords had made huge profits from the kelp trade, while keeping their tenants in grinding poverty. But as the lucrative industry declined, the lairds needed ready cash: something had to be found to take its place. The answer was sheep, Cheviots in particular.

The only problem was the tenants who occupied the crofts and scratched a modest living from land and sea. They had to go. But few landlords cared to dirty their hands with actually removing the people, preferring to sub-contract this work to their factors or agents.

One such was Hugh MacAskill, MacLeod's tacksman (defined by Dr Johnson as 'next in dignity to the laird') who seems to have taken to his duties with some energy and zeal. According to Neil Wilson, 'In total, some 250–300 people must have been evicted by MacAskill'.

Hardly a charitable man then, though a man of his time. By 1825 he was in possession of Talisker House and shortly afterwards, working with his brother Kenneth, he obtained the feu (effectively the lease) on a twenty-acre property at Carbost and, raising finance of some £3,000, built what we know today as Talisker distillery. Incidentally, while there are a number of ways of estimating what £3,000 in 1830 is 'worth' today, a reasonable case can be made for suggesting that it would then have had the economic power of almost £12 million.

Evidently then, the MacAskills were men of considerable substance and the distillery will have created some employment in the location. However, it seems that they then sought to exercise further control over their employees, many of whom were desperately looking for work as a consequence of the Clearances, being apparently reluctant to pay them in coin of the realm.

Two small curios survive in the safe at Talisker—small, privately struck coins representing payment for a half and one day of work at the distillery. They bear the legend 'Carbost Distillery' and 'Hugh MacAskill. Tallisker' (curiously, the distillery was then spelt with a double l). These are tokens to be used to purchase goods in MacAskill's company store, a system known as 'Truck'.

Essentially, employees paid in this way became indentured to their employer who, of course, was able to set the prices in the company store. Unscrupulous employers did so to their advantage, exploiting their employees and thus profiting twice from their labour. Though there were instances of benevolent employers who set fair prices so as not to take undue advantage of this captive market, the system was widely criticized for the widespread abuses associated with it. Even for Victorian capitalists the Truck system was unpalatable and much legislation was passed to restrict and later outlaw the practice, though it continued in some form late into the century.

Was Hugh MacAskill a benevolent or cynical employer or simply typical of his class and age? The archivists at Diageo suggest that he may have instituted the system to help control the supply of grain and that he is recorded as helping with poor relief in one or two cases. His record on the Clearances suggests otherwise, but probably the fairest verdict is that no one knows. Perhaps there was simply a coin shortage; such things did happen back then.

I do think it a shame that the tokens aren't displayed at the excellent visitor centre at Talisker. They represent a fascinating part of British social history which I suspect isn't well known or understood. The story is an interesting one and not entirely irrelevant to current employment issues—consider, for example, the debate on the rights and wrongs of zero-hours contracts.

It's not recorded how long the truck system was in operation at Talisker, and, in any event, the distillery was not a rip-roaring success for the brothers, though they expanded their holdings to some ninety-five acres, including the adjacent farm. When Kenneth died in 1854 the distillery, still then known as Carbost, was offered for sale in April the following year with an 'upset price' of £1,000—a significant loss on the original capital cost. This despite the claim by the auctioneer that the distillery was 'celebrated' and that for 'upwards of twenty years the quality of the Skye Whisky has been unrivalled'.

Not that they were the only ones to suffer losses. The whisky trade was experiencing severe disruption: consumption was declining and the large Lowland distilleries represented severe competition to rural operations, especially in Islay and

Campbeltown, where there were many closures. In some cases, the financial position of the failed ventures was so poor that it was not worthwhile for their creditors to pursue them.

Talisker at least survived. But from one family of energetic Skye capitalists I turn briefly to another and to Skye's second distillery at Torabhaig. This was the vision of Sir Iain Noble, who died in December 2010, so never saw his dream come to life. Shortly before his death I was considering suing him, so you will gather that I am not to be considered an admirer nor an entirely impartial witness.

Sir Iain had enjoyed considerable success in business, most notably as co-founder of the Edinburgh merchant bank Noble Grossart, and some two dozen other companies. In 1972 he purchased an extensive crofted estate on Skye and his interest in the island and the Gaelic language began to grow.

He established several businesses on his land and was instrumental in promoting Gaelic, most notably through the Gaelic college Sabhal Mòr Ostaig, which will be his most enduring legacy. More controversially, he was involved in granting land rights to the Skye Bridge and in 2003 attracted much opprobrium for a speech in Edinburgh denouncing English immigration into Gaelic-speaking regions, which some commentators thought racist or even Nazi in tone.

As the late Charles Kennedy, MP, commented at the time, with admirable understatement, 'Sir Iain is a well-known and colourful character'. He was certainly not one to disguise his views, nor was he afraid to pursue a course of action, no matter how unpopular or unfashionable, if he believed it to be correct.

Back in the autumn of 2006, I was asked by the Edinburgh architects Simpson & Brown to undertake some consulting work for Sir Iain on plans they had prepared for the conversion of his farm steading at Torabhaig to a small distillery. In particular, work was required on the proposed visitor centre and my consultancy practice was engaged to write a feasibility study and draw up some proposals for the interpretive design of the centre.

At first, all went well and our proposals and comments were well received. It then became clear that Sir Iain was fixed upon the idea of having a small 'black pot', such as would have been used by crofters and illicit distillers, on working display in the middle of the visitor centre. This was to be operated over an open fire, actually making whisky.

I pointed out that this was not only extremely unlikely to be approved by HMRC but that it also presented a fire risk and would never pass any health and safety inspection. He insisted that the proposals include such a detail, indeed that the still formed the centrepiece of the visitor centre and maintained, rather high-handedly I felt, that he would deal with the relevant authorities.

I declined to include the feature in our design and pointed out, probably quite pompously, that I was unable as a matter of professional reputation, not to mention the risk to my indemnity insurance, to include a proposal that I knew to be unacceptable. He insisted. I demurred.

Our increasingly fraught discussions continued for some time. Finally, I submitted a report which included the open fire and still, but made it abundantly clear that this was the client's direct instruction and had been included against our explicit recommendation.

Evidently Sir Iain, who had long dreamed of some plaid-clad peon tending the peat fire and watching the still whilst muttering incantations and spells in ancient Gaelic as boggle-eyed tourists looked on in wonder, and I shared a very different view of the role of the consultant. While I sought to offer expert advice, well-founded in best industry practice and having regard to law and practicality, my client evidently regarded me as little more than an amanuensis, on call to record his precious insights and gild them with the authority of independent endorsement.

Further discussion, by now quite frosty and distant, then ensued, with Sir Iain declining to settle our bill on the grounds that we had not understood his brief (we had, all too clearly) and the work was unsatisfactory. A stalemate ensued, but he had picked his ground well.

The sum involved was significant enough to be more than annoying but insufficiently large to call upon the services of m'learned friends, as no doubt he had calculated. While I considered my next step he unfortunately died, thus taking matters into the hands of a higher power than even the Scottish legal system. Our account was never settled.

I harboured severe doubts about the viability of this project, certainly under Sir Iain's quixotic guidance, and following his death frankly doubted if it would ever happen. Indeed, I said so in print and I maintain I was right — until, that is, the financial climate for small-scale farmhouse distillation changed very rapidly and the project was sold on by Sir Iain's estate to a properly managed and fully funded professional team.

So, as you read this, Torabhaig distillery is open and working. However, there is no open fire and working black pot to be seen. What is to be found is a remarkable blend of Sir Iain's thrawn determination to build something on a very challenging site and the practical skills and flexible engineering design of an experienced professional team backed up by a deep-pocketed owner. Sir Iain Noble could never have built Torabhaig, but without him Torabhaig could never have been built. It's a nice irony.

I was fortunate enough to see the very final stages of the installation work. The eight washbacks were in place, cleverly shoehorned into a cramped upper gallery. In order to fit them in, they are off-set, so one takes an unusual route through the space, twisting and turning as you move through to the tiny still house. It's an ingenious piece of design but I suggest you try to visit on a quiet day because, unless they set a punitively high admission charge, this is going to get uncomfortably crowded in peak season. I don't rate the chances of expansion here either; this is a unique operation that makes even diminutive Kilchoman look spacious.

But all credit to the installation team for working round the architect's onerous conservation requirements (Simpson & Brown are nothing if not meticulous and precise) and the absurd restriction of the space itself. Much of the credit for this presumably falls to the project manager, Chris Anderson, whom I knew when he worked for John Dewar & Sons and ran their five distilleries — so very different in scale and design.

Now retired, Chris was hired to project manage this installation. It is, I think it safe to say, not the distillery he would have designed. The building and its idiosyncrasies comes first here, and efficient design, ease of operation and smooth flow of liquid are a poor second. Despite the shortcomings of the layout, Chris has placed his considerable experience at the service of the tradition revived at Torabhaig and created a workable distillery out of an eccentric and very individualistic vision.

In fact, Torabhaig seems like a very indulgent project. The single pair of stills will produce 500,000 litres of spirit a year at most and the total cost of the project, including the necessary but expensive restoration of all the buildings and associated landscaping, must have exceeded £10 million, due in no small measure to the restoration costs associated with a historic, listed building.

That's a huge expenditure for such a modest level of production. It's not as if large numbers of visitors will generate income to offset the capital cost. Not only can the site simply not accommodate significant numbers of people, the proprietors seem indifferent to the possibilities of tourism and the design almost wilfully places the gift shop at the beginning of the tour. 'Exit via gift shop' is the mantra of the experience industry, for 'guests' may not be permitted to escape without the mandatory exposure to the crammed shelves of the inevitable retail space. It's vital to the economics of most similar operations and, indeed, so conditioned have we become to the idea of shopping as leisure that we're almost disappointed not to be thrust into a cornucopia of gleaming souvenirs and mementos.

So what's going on? On the face of it, Torabhaig is a UK-based initiative by the previously unheralded Mossburn Distillers, registered with Companies House as recently as August 2013 and with their head office listed as Jedburgh in the Borders. Though several projects are in gestation in the Borders it has not been a centre for distilling since the 1920s, and has never been regarded as a major regional power in the making of whisky.

So, again, what's going on? Mossburn Distillers is, it emerges, a subsidiary of Marussia Beverages BV, which ultimately belongs to a privately owned Swedish concern, Haydn Holding AB, which would seem to operate from a PO Box number in Malmö.

That leads us on a fascinating, if tangled, trail, for Haydn Holding, far from being some kind of tribute to the eighteenth-century Austrian composer and father of the symphony, Franz Joseph Haydn, is a diversified group with interests in publishing, property, wine production and drinks distribution, ultimately controlled by the Paulsen Family Foundation, a legal entity incorporated in Jersey.

At the head of all this is one seriously rich individual, Frederick Paulsen, Jr, a sixty-something Swede, said by *Forbes* magazine to be worth around £5 billion, due to his success in growing the family firm, Ferring Pharmaceuticals, a maker of speciality biotech drugs. Despite his great wealth, Paulsen maintains a low profile, spending most of his time, energy and money on philanthropic activities and some serious adventuring. This is a man who has descended 14,196 feet to the floor of the Arctic Ocean to find the 'true' North Pole (he's also the first human to visit all eight of the Earth's poles — don't ask me).

His life reads like a *Boy's Own Paper* adventure. Gasp as he pilots his microlight across the Bering Strait or to Lake Baikal; stand back to admire him digging for mammoths in Siberia, duly launching Mamont Siberian vodka in a mammoth-tusk-shaped bottle;[1] and gaze awestruck as he becomes one of the few Westerners to sail the 1,200-mile River Amur to the Sakhalin Gulf.

Along the way, he's funded fertility clinics in Russia; rat eradication in South Georgia; an art museum on the North Sea island of Föhr and a textile academy in Bhutan. Set against some of his multi-million-dollar philanthropic donations, a small farmhouse distillery on Skye looks like a minor, hobby project, though I am reliably assured that costs are carefully budgeted and strictly managed. Whatever your feelings on foreign ownership of the Scotch whisky industry, he is most avowedly one remarkable fellow.

As a result, Torabhaig is in an exceptionally privileged position. Most unusually for a new start-up, there are no founder's casks, bottling of the new make as 'spirit drink', crowdfunding or any of the normal cash-raising devices that we've grown so used to over the past few years. And perish the thought that there might ever be a Skye gin (though someone else is already planning a garage-style operation in Portree).

So far as the company have announced their plans, there will be nothing prior to a possible limited five-year-old release around 2022 and the standard expression will be a ten- or twelve-year-old, so we shouldn't expect any Torabhaig whisky to be freely available until 2027 at the earliest. Mind you, it will hardly be

1 The vodka seems okay if you like vodka. Can't see the point myself. But it is a rather striking bottle.

'freely' available; this is surely going to be a super-premium, luxury, boutique release. Presumably the owners hope and believe that the market for high-priced whisky will continue as strong as it is today. One can only wish them luck and admire their courage: it will undoubtedly be a very long time indeed before they see a return on their investment.

However, with Mossburn's immediate parent, Marussia Beverages, being in the business of alcohol distribution, and with companies in the UK, Europe and the USA, they should be as well placed as anyone to market the whisky when it finally matures; but perhaps the most remarkable part of the story is that Torabhaig is not their only project.

The hint is in the company's registration in Jedburgh, for there they have embarked on a very much larger project to build a hybrid distillery making both malt and single grain whisky. There will also be warehouses, a bottling hall, hospitality and conference facilities and the largest whisky shop in Scotland. Around 2.5 million litres of spirit will be produced annually, and the capital cost is reported to exceed £35 million. And while all this is going on they are also building a whisky distillery in Japan.

I left Torabhaig quite breathless at the audacity of the scheme. It is undoubtedly going to be exceptionally interesting; the restoration is flawless and the site outstandingly beautiful, with a long view to a small bay. I did approach Mr Paulsen's London representatives to see if he might grant me an interview, perhaps just a few minutes by phone, or simply a couple of questions by email. To be honest, I fully expected the answer to be 'no' (it's what I would have said myself to such impertinence), but wasn't really prepared for the long burst of sardonic laughter that followed. At least that was better than the usual PR obfuscation and evasions and, once my respondent recovered, he was kind enough to add 'we hardly know when we will see him ourselves'. As F. Scott

Fitzgerald once observed 'the very rich ... are different to you and me'.

What Sir Iain Noble, who by such standards was hardly even slightly rich, would make of it I do not know and don't care to imagine. I did notice, though, that his estate was valued at some £4.7 million, so he really could have paid me for my advice on his smuggler's still and the open fire, because I'm not rich at all.

As my wife and I drove back to Portree and a truly traditional Scottish hospitality experience, the sky seemed to blaze a re-markable shade of purple. That, and a reference in Martin Martin's 1703 classic *A Description of the Western Isles of Scotland*, was to help me create these pages.

Martin was a Skye man himself, hailing originally from Trot-ternish. In his great work he makes a number of geological obser-vations, including the note that 'stones of a purple colour flow down the rivulets here after great rains'. I hardly need remark that as a Skye man Martin would have had more than adequate experience of great rains.

Be that as it may, the drama of the Skye sky and the endorsement of Martin Martin (so good, you'll note, that they named him twice) convinced me that the only ink good enough for my Skye manuscript was an imperial purple of the highest quality. Once again, Pilot came to their rescue with their stunning *Yama-budo* shade in the peerless Iroshizuku range. It's available with two other handsome inks in an elegant presentation box that was so nice that I bought one for myself and one for a gift. This ink business was starting to prove something of an expensive affectation though the colour felt entirely appropriate for this somewhat overwritten passage (i.e. purple prose — see what I did there?).

But back to our traditional hospitality experience. With a booming tourist economy, Skye now features a number of interesting and different places to eat, even if your budget does not run to the famous Three Chimneys. Ours did not, so we set off to explore Portree, confident nonetheless of a better than average meal.

There is a cluster of small and mildly interesting-looking restaurants on Bosville Terrace so we made our way there. Most were closed but in one I could see a lady folding napkins. I knocked on the door. She made a 'we're closed' face, so I knocked again, at which she reluctantly opened the door enough for me to ask if a table might be reserved for that evening.

'We don't take bookings,' I was sharply told.

'Then how do I get a table?'

'You come and wait. What time did you want?'

'Seven thirty.'

'That's very popular. I've seen them queued onto the street.'

That didn't appeal, so I thought I'd ask again for a 7.30 p.m. table.

'I've told you,' she said, now very slowly and patiently as to an obtuse and particularly truculent child. 'We… don't… take… bookings.'

Memories of Scottish food service of old, not least in Port Ellen, came flooding back. Happily, we walked just a few yards along Bosville Terrace to the friendly Dulse & Brose restaurant, part of the Bosville hotel, where we secured the very last table. 'Dulse & Brose' for English readers means basically 'Seaweed & Porridge', but the food, and service, and I'm afraid, the price, was of a considerably higher standard than that might suggest.

As this isn't a restaurant review, all I will tell you is that we enjoyed it so much that we immediately rebooked for the following evening, when our experience was an even happier one. All is not lost for Scottish hospitality then, even if some old values live long.

I left my account of Talisker around 1854, when Kenneth MacAskill had just died and the distillery was offered for sale. The MacAskill brothers appeared to have lived in some style however, with Hugh the occupant of Talisker House, formerly the residence of the MacLeods of Talisker. It had been extended in 1775 but the house that we see today is much enlarged from the original, where James Boswell had stayed two years previously with Dr Johnson, noting a forecourt of 'blueish-grey pebbles' where he suggested that 'you walk as if upon cannonballs'.

Other notable visitors included the Swiss traveller L.A. Necker de Saussure who, in September 1822, had been handsomely entertained by MacLeod who, he relates, employed a piper to entertain them at dinner with pibrochs, the romantic arts of which 'for a long time resounded in the vaults of the castle of Talisker'. De Saussure thought it a fine house, still surrounded by the 'good many well-grown trees' that had earlier impressed James Boswell.

De Saussure evidently remained on Skye, living for many years at numbers 10 and 14 Bosville Terrace, where he died in November 1861. He was buried in Portree where his gravestone may be seen to this day. I assume that as he stayed for so long he was able to get a decent dinner on a regular basis—had he lived longer he certainly could have done, as Dulse & Brose trade from numbers 9–11. Tough to get a table when busy, though, and I don't suppose he cared for queuing in the street.

Strangely, Talisker and Talisker House pop up quite frequently in sci-fi and fantasy novels, not a genre that I had previously considered a likely source. Sparing you a lengthy exposition I was greatly amused to see in Brian N. Ball's *Planet Probability* (1973, but not a work that subsequently unduly taxed the printers) a reference to Talisker as a 'haunted, experimental planet [where]...the Alien that had supplied the Gene-key still lingered'.

Ball was evidently captivated by Talisker for it reappears in another of his long-lost efforts, *The Probability Man*, in which the planet Talisker is 'Pre-human'. I can't be certain, but I like to imagine that Mr Ball was an earnest student of the cask-strength versions.

I can't leave Talisker in the realm of the fantasy novel, however, without sharing my all-too-brief moment of rapture when I imagined that I had discovered Buffy the Vampire Slayer visiting Talisker House! I was more than unnaturally excited to read in the *Wisdom of War* that Buffy 'kicked in the doors' of Talisker House in her search for the Moruach Queen, but it turned out to be another Talisker House — testimony to the powerful appeal of the name in fiction. The Moruach, incidentally, is a sea creature from Irish folk legend, a sort of fairly benign mermaid with a taste for brandy, so it would be too fantastic to imagine her quaffing beakers of the peppery, smoky lava of the Cuillin.

Elsewhere, Talisker appears in a number of thriller and crime novels, generally to establish the manly characteristics of the pro-tagonist. Not for nothing are James Bond, 007 himself, and M seen drinking Talisker in *The World Is Not Enough* and *Die Another Day*. By *Skyfall* however, The Macallan seems to be the whisky of choice and Bond sticks with the brand in *Spectre*. Mind you, back in 1963 in the original novel, *On Her Majesty's Secret Service*, Jack Daniel's was his call. You would think he could make his mind up. I'm shaken, but not stirred.

The history of Talisker House is all very interesting, and the kind of digression that makes this journey so fascinating for me (I hope you're finding it worth your time, by the way), but it has long been detached from the distillery and so isn't exactly of contemporary relevance. But I like the way these things hang together and help us understand how the world once operated, beyond our own immediate experience.

As I was saying, Talisker distillery was on the market in 1854 and seems to have passed through various hands after it was eventually sold three years later for £500 (worth noting, by the by, that whisky hasn't always been a guarantee of riches, as that was only half the advertised price and a substantial loss on the capital cost). But by 1880, Roderick Kemp, a wine and spirit merchant from Aberdeen, had taken control, along with his partner, Alexander Allan. Allan was a part-owner at Glenlossie and a Morayshire legal officer, while Kemp was later to own Macallan.

There were evidently very different times for whisky, which was steadily recovering from the difficult trading conditions of the middle of the nineteenth century and heading, though no one knew this yet, towards what proved to be one of the industry's great periods of expansion in the century's final two decades.

While the whole of the industry was about to enter into something of a golden age, at Talisker something particularly special seems to have happened. Between 1857 and 1880 its reputation grew slowly. When Kenneth MacAskill died, the output was recorded as 500 gallons per week (note that the distillery would not then have operated year-round, as today) but was considered capable of producing up to 1,000 gallons 'with a little outlay in utensils' but no new buildings.

We know from Barnard that as late as 1886 the output was no more than 40,000 gallons annually. Compare this to nearby Tobermory, on Mull, of such contrasting fortunes, where Barnard records production of 62,000 gallons—more than 50% greater than Talisker.

Quite how and why it began to grow in fame isn't fully recorded or, if it is, I haven't been able to track it down, other than to conclude the reputation was based on quality and character, propagated by word of mouth, the press and popular culture. Perhaps the scattering of Skye folk through the Clearances and other migration was at least partially responsible for its growing status. Highland drovers—considered sound judges of whisky—were said to endorse it, according to the *Falkirk Herald* of January 1874 in a long tale of drovers extolling Talisker as 'a fine speerit', superior to brandy.

Clearly its fame had spread further afield, the London Evening Standard of 12 October 1872 referring to 'the famous Talisker distillery at Carbost', presumably confident that its metropolitan readers would find this quite familiar. For some time prior to this, the terms Carbost and Talisker seem to have been interchangeable but, by the final quarter of the nineteenth century, Carbost largely disappears and gives way to Talisker.

Talisker also features on occasion in the immensely popular novels of the prolific William Black, a Glasgow-born writer who at the height of his fame during the Victorian era was compared favourably to Anthony Trollope. According to his biographer, Wemyss Reid, by 1873 'it is no exaggeration to say that there was in England no more popular writer than Black'. For example, in his Stand Fast, Craig-Royston! (a three-volume work from 1891) the character MacVittie, an expatriate Scot and now a successful New York merchant, by way of demonstrating his social status addresses his guests thus:

'What would you like to drink, sir? I can give ye a choice of Talisker, Glenlivet, Long John, and Lagavulin; but perhaps ye would prefer something lighter in the middle of the day. I hope you don't object to the smell of the peats; we Scotch folk are rather fond of it.'

As is apparent from the above, Black was also a great fan of Lagavulin and a quotation from one of his more obscure works (which, believe me, is saying quite something) in praise of the distillery's water source still graces the front label of the official bottling of Lagavulin sixteen-year-old. It's probably the only place other than here where he's remembered; I doubt if anyone reads his novels today, which fell out of fashion soon after his death in December 1898.

But Talisker's big moment in the limelight came in 1887 when the renowned Robert Louis Stevenson described it as one of his three 'king o' drinks as I conceive it'. Now famous as Black was,

Stevenson's endorsement carried Talisker to a significantly greater audience. This was praise from a literary giant in the days before the dark arts of product placement and celebrity culture had taken hold of the popular imagination.

He was also partial to 'Isla or Glenlivet' and what's interesting about that is that he has nominated two regions, but only one distillery is singled out—a rare mark of distinction for the Skye brand. It's also worthwhile to note, and not widely remarked upon by those commentators that employ the quotation so freely, that it is from a poem entitled 'The Scotsman's Return from Abroad' which appeared in his anthology *Underwoods*.

Stevenson even goes on (it's a long poem that's frankly not to contemporary taste) to criticise the foreign whiskies he's tried 'frae Zanzibar to Alicante' (who knew?). These he 'abominates', and pours scorn on the 'foreign tricks an' pliskies'[2] of serving with lemon peel, ice and suchlike filth. That's all very well, and you might agree with his strictures, but it's a sort of Jekyll and Hyde performance.

However, the point is that by 1887, when Stevenson was approaching the height of his powers and already a considerable and well-known figure, out of all the whiskies of Scotland he picked Talisker to feature in the poem. It's a strange kind of tribute but, alongside the references in William Black's novels, one that marks Talisker's transition from a purely local hero to wider fame.

Whether or not it was because of this is not recorded, but around this date Kemp and Allan began to modernise and expand the distillery, though operations were greatly hampered by the lack of a pier to allow supplies to arrive easily and casks of whisky to be dispatched to customers. It was still necessary to float the casks out into the loch for them to be loaded onto a suitable boat.

Physical access was not the only challenge they faced. In 1894,

---

2 An old Scots term for a practical joke. Actually, quite a lot of Scotch whisky was served at this time as a toddy, with lemon, lump sugar and hot water. No ice, presumably.

R. Kemp & Co. were obliged to take legal action against a firm of London grain dealers, Hamlyn & Co., who had agreed to purchase dried grains made ready for them for shipment from Carbost. Hamlyn & Co. vs Talisker Distillery raised an important point regarding the legal principles of jurisdiction. The case eventually went to the House of Lords and was discussed in the *Harvard Law Review*. However, sensibly, the parties agreed to arbitration and the matter disappears from the record.

Here we are again, peering dimly into the viscometric whorls[3] of whisky's history. Perhaps Talisker gained some fame alongside Drambuie; as this hails originally from Broadford in Skye and was first recorded commercially in 1873 it would seem reasonable to assume that the original recipe for An Dram Buidheach (The Drink that Satisfies) involved a generous measure of the local malt. It certainly enjoyed local fame: in promotional material issued in the early 1900s by the short-lived Dailuaine-Talisker Distilleries Ltd it's claimed that 'To those who have ever travelled in the Western Highlands, the fame of Talisker will not be unknown'. I think there is some Scots reticence in this modest assertion, as clearly its reputation was more widespread that this implies.

So the taste and the quality of the whisky itself, spread through print media, has to be the principal reason why Talisker prospered as nearby Tobermory declined (especially as from 1916 onwards they were both, at least in part, absorbed into the ever-expanding DCL empire, with a full acquisition of Talisker in 1925). In earlier Dailuaine-Talisker Distilleries literature there is reference to a 'well-known sketch of Highland character' in which a brawny Scotchman, asked to drink brandy, replies, 'But to tell ye the truth I would rather whusky; the real Talisker is

---

3 My thanks to fellow whisky writer Mr Charles MacLean who, in the now sadly defunct *Scotch Whisky Review*, established the etymology of this most useful term, descriptive of the clouds which briefly appear in the liquid when water is added to a glass of cask-strength whisky.

the thing for me', going on to praise it as 'a fine speerit, the Talisker'. Yes, it's the story recycled by the *Falkirk Herald*, which some years previously had been published in *The Inverness Courier*.

Perhaps it was simply something to do with the whisky. Perhaps Talisker was simply better and those who liked its famously robust taste really preferred it to anything else. H.V. Morton's 1929 *In Search of Scotland* describes it as 'that remarkable drink which is made in the Isle of Skye and can be obtained even in its birthplace only with difficulty'. It is further described as 'a grand whisky!'

By 1930, in the first modern book about whisky and arguably the first ever written from the point of view of the consumer (the snappily titled *Whisky*), we have the great Aeneas MacDonald writing lyrically in praise of Talisker as 'the fine Skye whisky'.

He famously goes on to propose a list of the twelve most distinguished of Highland whiskies and suggests that his readers will think it desirable to settle for themselves whether Talisker or Clynelish takes up the final place. As he says, it is a problem to be decided only following 'prolonged spiritual wrestling and debate'—what today, perhaps, we would categorise as a #FirstWorldProblem.

There is, however, another problem, and that is that the whisky they are praising is not and cannot be the Talisker we find today. And why is that? Simply because until 1928 Talisker was triple distilled. As the Dailuaine-Talisker brochure confirms, '"Small stills" was the old maxim, and small stills reign at Talisker. There are six in number, arranged in three pairs of graduated size. In many distilleries the whisky is distilled twice, but at Talisker the process is repeated a third time, with the result that a very pure whisky is produced'. And here, at a time when the quality of many whiskies would have been variable, to put it kindly, we may see the key to Talisker's reputation for quality.

In that context, it's especially significant that Morton and

MacDonald were writing in an era when blends were pre-eminent, something which the latter energetically deplored. Only truly exceptional whiskies would then have been sold as 'self' or 'single' whiskies, as single malts were then styled. But what did the blending industry make of Talisker, with its pungent and forceful style?

A document survives from 1924, probably compiled for the great Edinburgh blenders George Ballantine & Sons, today part of Chivas Brothers. This classification of all the 131 whiskies then available to the blender (131 — such riches!) places Talisker fourth in the list of 'Crack Highlands. A'. Clearly Talisker from this period would have been keenly sought after by the brokers and blenders, sharp judges of a whisky's quality and not to be easily pleased.

But first Talisker had to achieve this eminence, which takes us back briefly to the Kemp and Allan era.

The partners had paid some £1,800 for the distillery and shortly began a programme of modernisation. This was financed, at least in part, by a Glasgow firm, Allan & Poynter, who also advanced money to the Strathclyde Distillery and Tobermory. Allan & Poynter, no relation to Alexander Allan, had grown up on the back of the expanding whisky business with warehouses in Glasgow and were possibly also blenders on their own account. Their extensive trade in importing wine and brandy had declined following the disastrous phylloxera infestation of European vineyards, and whisky, benefiting from cognac's fall, had taken up the greatest part of their business.

Shortly after this the owner of Allan & Poynter, a Mrs Scott, sold the company to her employees, James Whyte and Charles Mackay, who set up on their own account, rapidly developing their business as blenders. However, so far as may be determined, Whyte & Mackay were not major customers for the Talisker make.

As production grew under Kemp and Allan's experienced stewardship, the lack of a pier at Carbost became more and more of a problem and eventually Kemp grew frustrated with their land-lord, MacLeod of Dunvegan, and sold out to his partner. Shrewdly, he then went on to buy the Macallan distillery, and is thus closely associated with two of the greatest stories in malt whisky.

The pier was finally built in 1900 and the distillery further extended. By now, it had merged with Dailuaine and was effectively controlled by Thomas Mackenzie, a successful Highland distiller. But conditions had deteriorated in the whisky market after the Pattison's collapse[4] in 1898 and by 1915 the DCL were part-owners, taking full control in 1925. Triple distillation was abandoned shortly afterwards but, apart from temporary closure during the Second World War, the distillery went on quietly, producing mainly for the DCL blends, until 1960 when the still house was burnt down, resulting in a two-year closure.

Fortunately, in the rebuilding that followed, the distinctive external worm tub condensers were retained and the manner in which they cool the vapours from the five stills is thought to con-tribute to Talisker's distinctive fruity and peppery nose and taste.

Though some single malt whiskies from some distilleries were historically bottled by the DCL, supplies were notoriously intermittent and the various companies that made up the group-ing saw their branded blends as the absolute priority. That policy began to change as the success of Glenfiddich, Glenmorangie and The Macallan was eventually recognised and, in 1988, Talisker, then available in an eight-year-old expression, was included as one of the first of the Six Classic Malts, DCL's highly successful entry into this market. A small visitor centre was opened in the same year.

---

4 A notorious event in whisky history. The Leith firm of Pattison's were blenders and partners in a number of distilleries. The company rapidly expanded following a public offering of shares, with extravagant advertising and lavish personal expenditure by the owners, Robert and Walter Pattison. However, much of their trading proved fraudulent. The brothers were eventually jailed, but not before the failure of the company had bankrupted a number of other concerns and brought the late-Victorian whisky boom to a violent and juddering halt.

Since then, sales have grown and the range of whiskies has been greatly expanded. The visitor centre has been developed at a cost of £1 million, indicating the importance of tourism to the brand. Indeed, today the distillery can get very busy during the peak holiday season, with demand exceeding the available capacity, despite tours leaving every fifteen minutes.

Remarkably, the blenders have to get by today without access to Talisker, as the entire three million litres annual production is now reserved for sale as a single malt. It is still a fine speerit, a very fine spirit indeed — and none tastier than the cask-strength thirty-five-year-old expression that I tasted there under the watchful eye of the distillery's Stuart Harrington.

It's one of just 3,090 bottles (well, 3,089 now for sure) from the Special Release programme. It would have made a magnificent souvenir of our visit but I was forced to leave it on the shelf when Mrs B spotted the £850 price tag and most considerately pointed out that we didn't have nearly enough tokens.

So, with hardly a backwards glance, we kept on trucking and made our way off Skye.

Orkney

If stiff, wash with paraffin

I THOUGHT I'D SAVE ORKNEY UNTIL LAST. I never came here as a child, which was extraordinarily remiss of my parents, but I have come to love it. It's perhaps my favourite of the islands that I've visited, though I return happily enough to all of them and I'm aware that there's a lot of Orkney remaining for me to discover.

I also saved my Mont Blanc until last, largely because it comes with its own supply of little plastic ink cartridges. I don't normally like these. Though they're clean and quick and easy to use, and your fingers don't get inky while filling the pen, there is something of the schoolboy about them and I miss the intimacy of the ritual of refilling from an ink bottle, especially if I want to change the colour of the ink. A good twenty minutes of displacement activity can be involved in flushing out the reservoir of old ink, cleaning the nib, choosing the new colour, filling the pen, clearing the various bits and pieces safely away, and writing a few test lines just to be happy. Writers like displacement activity; it fills time as effectively as dissertations on ink bulk out their word count. (*Don't think we haven't noticed! Ed.*)

But observing that the ink was called Mystery Black made this an easy decision. I do find Orkney slightly mysterious—if not particularly black—so the choice was an automatic one. Mystery Black will reveal these secrets.

Incidentally, the Mont Blanc was another gift: it's a handsome and quite subtle model (not the vulgar bankers' ostentatious favourite, the Meisterstück 149 with gold detailing; perhaps the Donald Trump of fountain pens), but a substantial enough object regardless, which the Kavalan distillery in Taiwan very kindly gave me in acknowledgement of some work. They arranged to engrave my name on the barrel in an elegant script, presumably so that I could not sell it, but thankfully refrained from embossing it with their logo, so it remains possible to use it in decent company. It is at least a discreet reminder never to order shirts monogrammed with my initials.

On my most recent visits, I've been strangely drawn to Stromness. There's no distillery there today, but there was once, and it produced some of the prettiest promotional material of any distillery for any whisky that I've ever seen. One postcard in particular, showing a couple embracing and kissing in a rowing boat (in a perfectly chaste manner), is enormously charming. The young man, who is very properly dressed in a smart white shirt and dapper red tie, holds his sweetheart with one hand whilst prudently maintaining a manly grip on the oar with his left hand. Her lovely long white dress may be in some danger though, as it falls gracefully into the bottom of the boat where in my admittedly limited experience there is generally some slimy water slopping about, possibly with fish scales and other unmentionable detritus floating in it. I feel she may have some explaining to do when she gets home, though that would seem to be the very last thing on her mind at that moment.

As this scene is set in what I presume to be Stromness Harbour and the period is some time between the end of the First World War and the eventual closure of the distillery (the town, remarkably, voted in 1920 to go dry and that, at a time of general depression, was the beginning of the end for the little distillery), I assume their vessel to be locally made.[1] There was once a thriving boat-building community with a number of yards in Stromness, building skiffs, yoles and quills by traditional methods for the local market.

Though charming, innocent and even naïve we might find this image, I doubt that it would be permitted today. It would doubtless fall foul of the UK's stringent rules on alcohol advertising, quite possibly on two grounds. It might reasonably be argued that it links alcohol with irresponsible behaviour (not keeping a firm hand on both oars, the cad, and in the interests of health and safety both should be wearing life-jackets) and, even if that were blinked at, surely the image associates alcohol with sexual attractiveness. Such wickedness. Such sordid and debauched behaviour.

We live in such a humourless, literal age. If this advert were to be published today some grim-faced, self-appointed guardian of the public morals would fire off a condemnatory email to the Advertising Standards Authority, and probably the Portman Group as well, just to be sure. Their ponderous machinery would rumble into action; a stony-faced committee of grand industry panjandrums would sit in judgement and, in due course, their

---

1 In fact, I'm reliably informed that what we see in the picture—and all eyes are surely on this—is the forrard end of an undecked South Orkney Isles Yole. The giveaway is the broad beam and angle at which the forward strokes or planks have to converge to meet at the bow. Had it been smaller we would have been looking at a quill.

magisterial edict would be issued and the advert banned. The Stromness distillery and their agency would be named and shamed and a predictable storm of outrage would erupt on social media, thereby giving further visibility to the offending image.

But be honest. Are you offended? Do you feel inclined to take a young lady, unchaperoned, into the perils of the ocean? Would more young ladies fail to resist the siren call of the sea or their nautical beau if it appeared today? Do you now yearn for a dram of Old Orkney?

OO! Actually, I'd better stop here because yes I do. I'd very much like to try some Old Orkney from the Stromness distillery, though I'd imagine that they hoped to sell more than one bottle every ninety years or so. Still, their lovely advertising lives on as a memory of less puritanical times.

Stromness' most famous resident was the poet, author and dramatist George Mackay Brown, who lived there all his life. Being born in 1921 and living close to the site of the distillery he would have seen it work and die. Later, he came to enjoy alcohol and I like to imagine that some part of his upbringing in Stromness and proximity to the distillery's familiar spirit inspired his description of whisky: 'And the whisky, what is it', he wrote, 'but the earth's rich essence, a symbol of all fruit and corn and cheerfulness and kindling?'

A hundred of today's whisky writers, journalists and bloggers could toil for a hundred years and more and not better that.

Close to Mackay Brown's house and opposite the site of the distillery is the Stromness Museum, which is not to be missed. It is wonderfully old-fashioned, quite delightfully full of dead things in cases with no trendy interactive exhibits to amuse witless children. I loved it. You can spend several very happy hours here learning all about Stromness history, looking at more memorabilia from the distillery and even inspecting a set of gravity beads, the small glass bubbles that before the invention of the hydrometer were used by distillers to test the strength of their spirit.

However, on my most recent visit to Stromness, I was there to see some living history, courtesy of Scapa distillery. They have recently redesigned the label on their single malt to feature a silhouette of a yole. This, as we have learned, is a traditional wooden boat used by inshore fishermen, sometimes rigged for sail and sometimes set up as a rowing boat.

Having, appropriately enough, appropriated this particular piece of local culture, the Scapa marketing people felt impelled to 'put something back'. Quite what drives this, other than a vague sense of middle-class obligation, isn't clear, as distilleries, unlike, say, mines, don't actually take anything out of the locality, except perhaps water, of which there is no shortage; but let's gloss over that and simply recognise their noble intention to do a good deed.

The specific form that this takes is support of the Orkney Historic Boat Society, which exists — well, you can probably work out what it does. Neil Macdonald of Scapa's parent company, Chivas Brothers, was due to meet with the society and we went to see various of the fine old vessels, some in such disrepair that I would have thought them impossible to restore, that they are in the business of conserving and rebuilding. Even where a boat has gone beyond repair the pattern can be taken from it, I learned, and a new boat built to the authentic, original design, creating what one might think of as an Orcadian Ship of Theseus.

That, of course, proceeds from the presumption that there is someone capable of building a wooden boat by hand, using traditional methods and tools to create a wholly modern but genuine old boat. Once upon a time, boatbuilding thrived in Stromness but, one by one, the tiny yards closed down, the craftsmen died and the skills were lost. Today, only one yard survives with just one man working there, keeping the old ways alive, so naturally I went to meet him.

His name is Ian Richardson and, to one whose hands are soft

and who would struggle to tell a hammer apart from a screw-driver (that would be me), his boatyard is a place of wonder. It is draughty; it leaks and it is scattered—apparently at random—with mysterious bits of wood, bits of boat, bits of everything and anything which, if questioned, Ian will unhesitatingly identify as if it were obvious to anyone what they are and what they are for. Trust me on this, they're not. He was immensely patient and courteous with my ignorant and no doubt ill-informed questions and I came away with great respect and admiration for his work.

The Orkney Historic Boat Society work with him on restoration, and with growing interest in traditional boat building and support from the distillery he has been able to take on a trainee to ensure that his knowledge will be passed on to a new generation. We discussed the price of a new boat but it soon became clear that it was beyond a whisky writer's budget, quite leaving aside the fact that I live miles from the sea. I did discover however that someone famous had bought one of his boats; I seem to remember it was Sir Cameron Mackintosh, or perhaps Sir Elton John. To be completely honest, I wasn't listening too closely at that point, having become oddly fascinated by a strangely shaped bit of wood and trying to decide whether it was a vital part or an offcut. These things matter, or at least they seem to matter at the time and afterwards you're too embarrassed to ask was that Cameron Mackintosh or Elton John. Let's settle on Cameron Mackintosh; he seems a more likely boat owner than Sir Elton.

At this point I was supposed to attempt to sail a restored yole around Stromness Harbour with the ever affable Neil Mac-Donald. But the truth was that it was cold, raining very heavily, gloomy and overcast so we thought better of it. Good-looking though Neil is we could not really recreate that Old Orkney postcard anyway so we decided to go and look at some more boats.

These included the exceptionally stylish *Esmeralda*, an elegant racing yacht designed by Uffa Fox. I was quite proud to recognise his name as he is undoubtedly the only boat designer I could recall, even at peril of my life.

Also on show was the reconstructed Neolithic boat which featured in the BBC2 programme *Britain's Ancient Capital*. This was constructed using hides stretched over a simple wooden frame. A crew of hardy volunteers then rowed it across from Orkney to the mainland. However accurate the re-creation, and I don't suppose anyone can ever say with certainty, I later thought about this flimsy little craft when standing at Port Askaig on Islay, watching the tide race down the Sound of Islay and thinking about the Neolithic travellers who had settled at Rubha Port an t-Seilich. I've landed at Port Askaig when the CalMac ferry has had trouble with the tides; quite what it must have been like 12,000 years ago, in a fragile skin-and-timber-frame construction without the RNLI's *Helmut Schroder of Dunlossit II* on standby I cannot imagine.

The trip to Ian Richardson's yard and to the various historic boats scattered around Stromness and the immediate area was linked to a visit to Scapa distillery. There are some fine old dunnage warehouses at Scapa, but sadly these are no longer considered safe for the storage of casks of whisky. To avoid them lying idle and possibly deteriorating, the plan is to make them watertight and allow the Orkney Historic Boat Society to use them for storage and as a workspace for restoration projects. What an admirable and immensely practical form of support. Perhaps one day they can even be open to the public, provided that the usual sorts of liability issues and tangled health and safety problems can be overcome.

The Scapa distillery, as you may have guessed from the name, lies right on the shores of Scapa Flow, a body of water which will be forever famous for the various naval ships that lie wrecked there. It contains, apparently, just under one billion cubic metres of water which is, in layman's terms, a lot—more than enough to conceal the fifty-two ships of the German Navy's High Seas Fleet which were scuttled there on 21st June 1919. With nine German sailors shot as they attempted to sink their ship, the entire operation, prepared in great secrecy, was managed with minimal loss of life. Most were eventually salvaged; both quite remarkable achievements in their different ways.

That is more than can be said for the other great naval wreck in Scapa Flow, the British battleship HMS *Royal Oak*, which was torpedoed by the German submarine U-47 in October 1939. Out of a crew of 1,400 men, some 833 died and the wreck is now a protected war grave. On occasion, a slick of oil from the *Royal Oak* may still be seen on the surface of the water.

Some years ago, I had the pleasure of accompanying a Russian submarine admiral around Scotland and we were entertained, most generously, by Chivas Brothers in a handsome house they own in Keith. Naturally, there is a bar and we retired there after dinner, where the admiral spotted a bottle of Scapa.

His eyes lit up and he grew very animated. He claimed to have once clandestinely taken a Soviet submarine into Scapa Flow to re-enact the daring raid of Kapitänleutnant Günther Prien, commander of the U-boat which made the successful attack on the *Royal Oak*. He was full of admiration for Prien's audacity and seamanship (and, I think, his own). Mind you the Admiral, who was a great fan of Ardbeg, was also full of Chivas' whisky at the time and may have embellished his salty old mariner's tale for the sake of his awestruck audience. As a footnote to this story I would advise you, if ever you fall into the company of retired Russian admirals, not to attempt to match them drink for drink and, whatever else you do, do not get involved in competitive toasting. It will end badly, especially as they may well have command of nuclear weapons and you certainly do not. At the very least, this gives them bragging rights.

In certain conditions of light and weather Scapa Flow can look foreboding and gloomy. It is easy to believe the old story that it was cursed by a Shetland woman who lost a thimble there and proclaimed that whales would never again be caught in Scapa Flow until it was found. It seems an extravagant penalty but as today the whales would be protected and shown great reverence, the curse may be considered to have expired.

Today the Scapa distillery likes to make much of the story, found in James Wallace's 1688 work *An Account of the Islands of Orkney*, of an ancient tradition that a large cup was kept at Scapa, which, on the arrival of a new bishop, was filled with strong ale and presented to the divine. His countenance was closely watched and if he happened to drink it cheerfully, the locals considered this a favourable omen. This, they concluded, was a noble bishop, likely to serve them for a good many years. Quite what happened if they encountered a grumpy or, worse still, a teetotal cleric is not related.

Sadly, Wallace concludes that the tale is 'still believed here and talk'd of as a truth, though now there be nothing in it'. Having said that, I'm assured by the friendly visitor centre staff that there is a welcome dram to be had by any thirsty churchmen calling at Scapa distillery.

At least today they will receive a friendly greeting. For many years, the distillery was closed to the public, despite the obvious success of the rival Highland Park visitor centre, and would-be visitors were turned away. Like so many island distilleries, Scapa has had a chequered history and has flirted with closure more than once. As recently as ten years ago, you could have got short odds on its closure, yet it was revived by its then owner, to general surprise, and then acquired by Chivas Brothers.

This has been its great good fortune. Unlike their great rivals, Diageo, Chivas do not own another island distillery and so were able to concentrate on marketing and promoting this one in quite a single-minded fashion. They have invested in the stocks necessary to build the brand; improved the packaging and opened a small visitor centre.

All of this has proved timely. Not only have island whiskies become more fashionable and thus sought-after, but Orkney itself is experiencing something of a tourist boom, largely owing to the increasing number of large cruise ships now calling at Kirkwall. Their passengers are naturally anxious for some shore-based excursions and, for those not attracted by the charms of Orkney's history, a distillery visit is an appealing option.

Highland Park has proved particularly popular with the cruise-ship market, but this has had the unexpected consequence of deflecting other visitors to the island to Scapa, which is also attracting those whisky enthusiasts who have previously visited the larger distillery and are now looking for a more boutique alternative.

Elsewhere the effect has been dramatic. Having not visited Kirkwall for some years, I was unprepared for the impact of the affluent cruise passengers on the main shopping street. Where previously there had been empty units and a sadly run-down air, Kirkwall's main thoroughfares now look sleek and prosperous, fitted with every variety of gift emporium you can imagine. The island's long-standing craft community, noted especially for knitwear and jewellery, is presumably overjoyed by this unexpected bounty from the sea. The shops themselves seemed busy, but there seemed a strangely uniform quality to the offer, until I realised that products must be, above all, expensive and small.

Size matters here because there is limited space in a cabin, however luxurious, and also because many of the cruise operators have a prohibition on bringing alcohol on board, as for some strange reason they prefer passengers to buy from the ship's bars. A buoyant trade has sprung up, therefore, in smaller sized bottles which can more easily be smuggled aboard in a handbag and stowed in cabin baggage. The shrewd off-licence operators of Kirkwall have observed this particular market requirement and stocked their shelves accordingly in their anxiety not to deny the passengers their clandestine souvenir of life on dry land.

Scapa's visitor facilities extend to a curious wooden structure on the lawn at the back of the distillery, facing Scapa Flow. It resembles a giant foghorn, or megaphone, and was constructed, or so I was assured, from the timbers of an old washback, which accounts for the curious patterns on the wood. It is sufficiently large for a grown man to stand inside the open end, but according to my guide has no function whatsoever apart from being an intriguing conversation piece. Having said that, it is amusing to walk around and into it and speculate on what it might mean were it an exhibit in contention for the Turner Prize. It is then even more amusing to agree that it is simply a folly, with no greater intent than to make us feel better about ourselves. I would defy anyone not to leave without a silly grin on their face, and while this is a normal enough condition amongst visitors departing from a distillery, the wooden foghorn achieves this without the aid of whisky.

The distillery was built in 1885 as part of the general boom in distilling and to supply the flourishing blended whisky market at the end of the Victorian era. Eventually it ended up owned by Allied Distillers who sadly neglected it and allowed the buildings to deteriorate. There was no full-time distillery workforce and the only production after 1994 was carried out by a team contracted from Highland Park, presumably mainly to ensure the equipment did not seize up or to check that it had not been vandalised or the copper stolen.

By August 2005 it was in a poor condition and more likely than not was on the verge of permanent closure. The wiring had been condemned, the roof of the main building above the mash tun had failed and the still room, with its dramatic picture window, was home to a small bird colony (species unrecorded). The warehouses had been emptied and the offices abandoned. There were, realistically, only two options: closure or total refurbishment.

To their credit Allied Distillers, never noted for their commitment to or understanding of single malt whisky, opted to bring the old place back to life and, for an investment exceeding £2 million, that's what they did. Ironically, within a little less than a year, Allied was broken up and Scapa was acquired by Chivas Brothers, who knew a good thing as soon as they saw it.

They continued the refurbishment and investment programme and today the whole place is unrecognisable from fifteen years ago. The distillery is relatively compact, with just one pair of stills, capable of an output of around one million litres of alcohol annually. For the casual or uninitiated visitor, it is an easy-to-absorb site that offers a quick and simple introduction to whisky distilling in a dramatic setting.

The enthusiast, however, is here with a more arcane motive, for Scapa presents the opportunity to see a Lomond still in action which, you will recall, is the type of still used at Bruichladdich to produce The Botanist gin. Apart from those at the Loch Lomond distillery (which isn't open to the public), it's the only Lomond still left that is still making whisky, which makes it of considerable historic interest.

I'm not convinced, however. The original Lomond stills were designed to be used as the spirit still, thus having the greatest influence on the final spirit character, and essential to their operation were the rectifying plates located in the tall, cylindrical neck. So, there are two problems with the Scapa 'Lomond' still. First it works as the wash still, not at all what the designers originally intended, and second the plates in the neck have long since been removed as they proved troublesome in operation.

What we have here then, I maintain, interesting though it may be to the dedicated whisky enthusiast, is a notably flat wash still with a wide and tall cylindrical neck. With the plates removed, the potential to alter the reflux and thus the spirit character has been lost. In fairness, though, the neck is so large that there will be unusual amounts of copper contact and, in addition, a purifier pipe (similar to Ardbeg) contributes to the reflux that creates the fruity and oily spirit.

And, also in fairness, that end result is really what counts. Here Scapa does deliver a very pleasant, medium-bodied, fruity whisky that is undemanding, self-effacing and easy to drink, whether in the classic Skiren release or the newer, slightly peated Glansa version.

The revival of Scapa has, quite possibly, been even more successful than its owners anticipated. The result is that stocks of mature whisky which, due to the erratic production, were never extensive have been largely exhausted. There was a fourteen-year-old style, then a sixteen-year-old bottling, but both of these have long been withdrawn. As a result of the need to stretch out the available whisky, both Siren and Glansa are bottled at 40% abv as opposed to anything more robust, and neither carries an age statement. While hardcore enthusiasts might be troubled by this, it did not seem to deter visitors to the distillery who were happily removing these from the shelves as I enjoyed a reflective dram.

Occasionally there are special editions, such as the Scapa Jutland sixteen-year-old, released in June 2016 to mark the centenary of the First World War Battle of Jutland and raise funds for the RNLI. Just 249 bottles were available, at £200 each, and canny islanders queued from early in the morning of the day of release to snap up a bottle. Many soon featured on auction sites, where the price soared to over £600 before the auctioneers' commission.

That's a handsome profit for getting up early one morning with an anorak, flask of tea and a credit card. I can see why the distillers lose patience with this sort of profiteering though: they have made the whisky, watched over it for sixteen years, bottled it, packaged it, paid duty and VAT and donated their profit to charity, only to see carpetbaggers exploit the situation and pocket a gain far in excess of that recorded by the people doing all the hard work. It must be exceptionally frustrating and yet all they can do is raise the retail price even further and be greeted with howls of outrage from their regular drinkers and the usual social media mob, who like nothing better than to share their outrage. In fact, it's just occurred to me that on a £600 sale, the auctioneer makes more money than the distiller. And that is outrageous.

I can tell you that it is a most enjoyable whisky, however. The kind people from Scapa bought me a measure at the Lynnfield Hotel's Whisky Bar where there is a bewildering choice of old and rare drams. I enjoyed it, the company and the setting, though I don't imagine I would have paid £600 for it. However, at least that bottle was opened and being drunk so I felt I had done my duty by it.

With the whisky commemorating the Battle of Jutland, naval heroism came to mind and, quite unbidden, some verses swam out of my distant childhood to the top of my fevered memory. I promptly attempted a parody which, as far as I recall, went like this:

> The boy stood in the whisky bar
> Whence all but he had fled
> The glass he held was full of tar
> He looked on it with dread.

That's terribly unfair on Felicia Dorothea Hemans, author of the ballad 'Casabianca' that was so widely lampooned when I was a child, and indeed Scapa Jutland, which is not in the slightest a peaty, smoky whisky. As you're highly unlikely ever to taste it, I should put on record that it's quite sweet and fruity with hints of vanilla and butterscotch. Quite delicious, in fact, and well worth tracking down if you could ever find a dram at anything approaching a reasonable price.

Kirkwall being small, and the Lynnfield quite a place of pilgrimage for whisky fans, our party inevitably bumped into some other distillery people. If you like looking at old bottles, and gasping at the price per dram, or simply studying loads of whisky memorabilia, then you can spend many happy hours here. Along with the array of bottles that fill shelves, bookcases and cabinets there's old packaging, bits and pieces of redundant equipment and some more Old Orkney advertising. Given how long their marketing material has survived, it's both sad and poignant that the distillery, once known as the Man o' Hoy, never made it past the grim 1920s.

But never mind that. Distilling on Orkney is evidently flourishing today. It's not simply whisky, however, for Orkney is as caught up in the gin boom as anyone else. As I write, there are actually three operations and there was quite a race to be the first Orkney gin onto the market.

Discerning Orcadians are now confronted with the paradox of choice: like buses in the rain, either there are none or they all come in a rush. On the one hand, they could try Johnsmas or Mikkelmas Orkney Gin from the Orkney Gin Company, which is a small family operation on the island of Burray, between South Ronaldsay and the Orkney mainland. They unashamedly style it 'bathtub gin'. Or very soon, if all goes to plan, the Deerness distillery will open their doors and their gin will be available (the building looked finished, though starkly agricultural, as this went to press).

But, so far, the crown must surely go to the Orkney Distilling Company who, by now, are producing their Kirkjuvagr gin (say it, kirk-u-vaar) in a brand-new distillery in the very centre of Kirkwall.

The business is the brainchild of Stephen Kemp and his wife. Kemp's family run a successful, and busy, local building company and happily for this project also owned some run-down and dilapidated warehouses very close to the main harbour. This, of course, is where all those cruise ships dock and their passengers disembark, thirsty to experience all that Orkney can offer.

When I went to meet him I, rather bluntly, expected a small, back-of-the-garage type operation with one or two of the tiny Portuguese stills that you can buy off the web. Far from it: this is an impressive, well-thought-through business that aims to build a solid base on local support, with the tourist trade providing the opportunity to carry the brand internationally. The stills are Portuguese, in fact, but decent-sized ones from Hoga, not the ones you can pop on the cooker at home.

Even before meeting Stephen Kemp I kept seeing his bottles in every bar that I called into (I tend to go into a lot of bars; I call it research), and the staff spoke enthusiastically of the product. That kind of trade support is vital and only built up by hard work and personal contact, so I was impressed by the effort that had obviously gone into this even before trying the product.

The first batches of Kirkjuvagr gin were made by the Strathearn distillery, near Perth, who helped Stephen develop and perfect the recipe. The key to the taste, as ever, is the use of local botanicals. Here a variety of angelica, first brought to the islands by Norsemen, is added to locally grown roses, borage and some Orkney bere barley.

With the distillery now in operation, good local support, more than a hundred cruise ships a year and Orkney's established tourism business, I fully expect Kirkjuvagr gin to prosper and I very much look forward to returning to see the distillery at work. Hopefully, all three Orkney gins can coexist and develop their respective businesses to the greater good of all.

I touched on bere barley earlier, while discussing Arran's one effort with this grain. It's also employed by Bruichladdich, who are the principal champions of this ancient variety and, on a previous visit to Orkney, I had been with my wife to the Barony Mill to see it being milled into the flour for bere bannocks.

Today, this is the sort of 'artisanal' product that you buy at fashionable farmers' markets, paying an outrageous premium over the price for any sort of a supermarket equivalent, but walking away in a cloud of moral superiority, buoyed up by your new-found status as an ethical shopper. Served with a meal in a high-priced restaurant, or tried occasionally as a local speciality, they're fine. But they're pretty hard going, if one is completely honest, and if they were all you had all year round, perhaps with a little porridge for variety, I imagine you'd pretty soon tire of them.

Anyway, we bought some and pretty soon tired of them. The flour itself was quite hard work to cook with and I began to see why bere had fallen out of favour. But the Barony Mill has a larger part in this story, for it's here that most of the bere that's grown today is milled. But how and where is it grown? And why?

To answer those questions, I went to see Dr Peter Martin at the Agronomy Institute in Kirkwall. He explained that bere is what's known as a 'landrace'; essentially, if I understood him correctly, a very ancient crop variety that has not been selected by scientific plant breeding but evolved out of traditional farming practice and the unconscious efforts of many generations of farmers. Peter went on to inform me that up until the end of the nineteenth century most farming in Western Europe was of landrace varieties which had evolved to meet specific local conditions, and that hybrid barley strains, bred for easier growth and higher yields, were not introduced until the early part of the twentieth century.

Most varieties then slowly fell out of favour. Bere, which was well adapted to the growing conditions of the Scottish Highlands and Islands, clung on longer. However, it was gradually replaced on the mainland and found only on smaller, old-fashioned farms, largely on the islands, notably in Orkney where local demand for the bere bannock ensured that it had some sort of economic viability.

Back in 2002, the Agronomy Institute were challenged to come up with some higher-value uses for bere, the thinking being that this would ensure the long-term survival of the variety. Given that bere was first mentioned in print in English in the 1530s, but genetically can be traced back to very early crops, perhaps as long ago as 3000 BC, it seemed to the scientists that it would be a small environmental tragedy if it were to be entirely lost to modern agriculture.

In fact, it has several advantages as a crop: it is low input (that is to say it does not require much fertiliser); it does not need fungicides; and, having a short growing season and a tolerance of high winds, it is well suited to northern Scotland. Tasty whisky also results, which some may consider an advantage.

The institute, logically enough, hit on the idea of reintroducing bere to the brewing and distilling industries and, slowly but with growing success, they have interested a number of smaller producers in its potential. As products made with bere are able to command a premium, so the interest of the farming community has grown, though the institute is still a major producer.

Several farms on Orkney are now growing for Bruichladdich, initial trials on Islay having failed. Bruichladdich's Bere Barley 2009 release comes from four different Orkney locations and, with their interest in provenance and traceability, the farms are identified on the outer packaging and the label: Weyland and Watersfield; Richmond Villa; Quoyberstane; and Northfield. All of the farmers work on the basis of minimum intervention and are concerned with a low-input, low-intensity, low-impact approach, treading lightly on soil which may have been in their families for generations.

This grain represents the authentic roots of Scotch whisky; a very real link to the farm distillers of the earliest records. While modern varieties are easier to grow, more profitable, higher yielding and easier to process through the distillery, the growing of bere and its use in whisky represent a rejection of many contemporary values and a conscious decision to engage with the landscape in a fundamentally different way from the modern agri-businesses that dominate cereal production.

For that reason, I was interested that Stephen Kemp had incorporated it into his gin and was looking to grow bere on land owned by his family. And I was even more interested to learn that the islands largest distillery, Highland Park, had been running trials of another heritage variety, Tartan, since 2009.

Varieties of barley go in and out of favour, partly due to the demands of the brewing and distilling industries and partly due to simple economics. For much of the nineteenth century, Chevallier[2] was the dominant strain, according to some sources accounting for as much as 90% of the barley grown in Britain by the 1880s. It then gave way to Plumage Archer, which was preferred by the brewing industry. Immediately prior to the Second World War, Chevallier accounted for less than 2% of the crop and was more or less obsolete—newer, higher yielding varieties having displaced it.

By the end of the nineteenth century, however, the Scotch whisky industry was importing grain from outside the UK. As just one example, the Barley Delivery book for 1893 for Glenglassaugh, near Portsoy, records that the distillery obtained around one third of its barley from Russia. Despite the cost of transport, it was still significantly cheaper than Scottish barley purchased from neighbouring farms. The distillery was paying up to 27/- per quarter for local, Scottish grain compared to 22/3 per quarter for the Russian cereal. Glenglassaugh were not alone; other distilleries also sourced on the international market.

2 Not a spelling error. It's named for the Reverend Dr John 'Barley' Chevallier of Aspall Hall, near Debenham in Suffolk, who helped develop it in the 1820s.

Interestingly, Chevallier has now enjoyed a small-scale revival as a 'heritage' grain and is used to produce traditional 'real ale' by the tiny Stumptail brewery in Norfolk. But, in its pursuit of greater yield at ever lower cost, the brewing and Scotch whisky industries have used a number of varieties—a sort of cereal arms race if you will—with each innovation being replaced in turn ever more rapidly.

Chevallier gave way, as we have seen, to Plumage Archer which, in turn, was toppled by Proctor and Morris Otter, only for these varieties to be replaced by Golden Promise. This was once widely used; one famous single malt going so far as to describe it as one of the 'Six Pillars' of their brand, a claim which has since been quietly dropped (marketing messages having an even shorter lifespan than barley varieties). Golden Promise enjoyed a good run, however, but by the early 1990s, Triumph was preferred, only to be quickly overtaken by Chariot and Chariot Harvest. Today, Optic and Concerto are favoured.

This is all about yield or, to put it another way, money. Poor old Chevallier gave around 300 litres of pure alcohol per tonne of dry malt; today the distiller (or his accountant) expects to get 450 litres or more from the same weight of malt. In a world where blended brands predominated and there was much competition on price, this mattered. Minor variations in flavour would be masked by the impact of the cask or smoothed out in the blending process and increasingly the drinker was assured that barley variety made little or no difference to flavour.

Well, if you have had the opportunity to compare new-make spirit from different varieties of barley but distilled on the same plant under the same conditions, you will know that this is not so. There are differences and, in the new make, they are quite apparent. Whether or not the distiller wishes to preserve those differences through the maturation process, so far as is possible, is a matter of taste and fashion for the most part. Current practice, especially for malt whisky, is to give greater weight and emphasis to the contribution of the cask. Currently, this is said by many distillers to account for 60% or more of the flavour of the bottled whisky.

However, I have sampled alternative barley varieties as new-make spirit in both Scotland and Ireland and the differences between them were both obvious and immediately apparent, even between the same strain grown on different farms. Whether you prefer one sample to another is only a matter of personal preference; what is, I think, undeniable is that there exists a perceptible difference. Quite why large parts of the industry persist in maintaining that no difference exists baffles me. This, of course, lies at the heart of all Bruichladdich's noisy protestations about terroir and the homogenisation of much whisky flavour.

So with that in mind, the fact that Highland Park is quietly experimenting with Tartan is exciting news.

Highland Park is the oldest distillery on Orkney, dating back to 1798. At least that is what the elaborate wrought-iron sign above the distillery entrance proudly proclaims, and who am I to quibble with a history wrought in iron? It certainly feels old as you step inside the cramped courtyard, with the stills hot and busy just a few yards to your left.

There are lots of stories about the foundation of this most atmospheric place. The distillery's favourite, and the one most widely told, concerns the activities of one Magnus Eunson, churchman and bootleg distiller, who was said to hide casks of new-make whisky from the prying eyes of the Excise officers by concealing them under his pulpit. To add verisimilitude to this particular narrative, legend relates that in the presence of the government men the pious Eunson would 'announce the psalms in tones of unusual unction'.

Other accounts have it that a farmer named David Robertson

founded the distillery and that it has alternatively been known as Rosebank or Kirkwall. I will choose to believe in Eunson, the brewing beadle, clandestine cleric and viscometric verger. Apparently, he was also a butcher some of the time. The gig economy isn't all that new it would seem.

Whatever the truth, the whisky made here has long enjoyed a high reputation and enjoyed early fame as a single whisky. The King of Denmark shared a bottle with the Emperor of Russia on board Sir Donald Currie's yacht in 1883 and they apparently concluded that it was 'the finest they had ever tasted'. Of course, we don't know if they really knew the first thing about whisky, but a little royal patronage never hurt anyone.

The renowned London wine merchants Berry Bros & Rudd bottled Highland Park under their own marque in the early part of the twentieth century and a small cache of bottles labelled Highland Park Reserve 1902 survived and were eventually confirmed as genuine by carbon dating conducted at Oxford University. Not all the old bottles of single malt suddenly emerging on auction sites are genuine, and there have been some scandals, so such verification is highly significant. Even distilleries themselves have been fooled: back in 2004 Macallan were embarrassed when it was revealed that around a dozen of the expensive antique bottles they had bought for their corporate collection were modern fakes.

At the same time, though, the greatest part of Highland Park's output would have been used in blending. Back in 2010, in an Islay bed and breakfast, I was lucky enough to stumble upon a promotional booklet for the distillery's new make for the 1924/25 season. It was crammed onto the shelves alongside some faded Hammond Innes paperback, ancient *Good Food Guides* and tired Christmas *Beano* albums. There it sat, abandoned and forlorn, perhaps carelessly discarded by an earlier guest (a blender, I'd like to imagine, in search of the island's pungent spirit) and of no consequence to anyone—until of course, I expressed an interest in buying it, when it suddenly acquired a remarkable rarity value.

However, its very obscurity had preserved it from a jumble sale or house clearance so I regarded it as a noble mission to save this modest publication. And I was very glad that I did because it emerged that not only was 1924 an important year for Highland Park, but it also turned out that the distillery did not know of the book and didn't have their own copy. And what is more, on reading the book, happily titled A Good Foundation, it seemed there was something to be learned about Highland Park's twentieth-century marketing. All in all, it was worth picking over the shelves of that B&B, looking for something to kill an idle half hour.

By coincidence, 1924 was the year George Ballantine & Sons drew up their list of whiskies, ranking them for their value to the blender. Highland Park was, they thought, amongst the very finest: they rated it one of the elite 'Crack Highlands. A'. High praise from one of the most renowned and long-lived of blending houses. So presumably recipients of A Good Foundation, hard-nosed whisky brokers and blenders, were well aware of the qualities that Highland Park could bring to their products.

What is valuable today about the book are the photographs of the distillery and the description of the processes, which we can compare to present-day practice. In their marketing, Highland Park make much of what they term 'five keystones': hand-turned malt; aromatic peat; even-paced cool maturation; sherry oak casks; and cask harmonisation. With the exception of that final rubric, all may be found in the 1924 pamphlet.

I enjoyed taking the little pamphlet to the Highland Park people, I enjoyed their pleasure (and surprise) in observing that their twentieth-century marketing was rooted in some demonstrable truths and I particularly enjoyed reprinting it for them to use, once again and more than eighty years on, as a promotional tool, but this time for an educated consumer, not a trade buyer. Too often, marketing is smoke and mirrors, or a tottering edifice built on a slender foundation of truth, so it was gratifying to see the link proven over eight decades.

However, it helped that Highland Park is one of my favourite distilleries to visit and one of my favourite whiskies to drink. With a very few exceptions (the excessively sherried Dark Origins release comes unbidden to mind here) I have enjoyed almost everything I've tasted from Highland Park, almost to the extent that I completely forgot to look at the price tag when once presented with their sublime fifty-year-old release.

As a place to visit, it's hard to beat, not least because it's on Orkney. That means that you have made something of an effort to get here and will be anticipating the visit with more than the usual fervour, and, unless a cruise ship is in port, you will likely be in the company of fellow enthusiasts. In fact, so enthusiastic are some of the more dedicated followers that they happily pay £1,000 each for a very special, extended visit during which they get to actually work at the distillery. I've never gone that far, though I believe they customise your overalls and boots and you get to take them home with you!

One of the first jobs given to the visitors with too much money for their own good is to try cutting some peat at the distillery's peat banks at Hobbister. They have several thousand acres of land here, alongside a bird reserve maintained by the RSPB, home to hen harriers, grouse, red-throated divers and curlews who seem largely untroubled by the activities of the peat cutters working nearby.

Orkney peat is different in character to that found on the mainland of Scotland or on Islay, due to the different vegetation and the island's climate. The distillery also go to the trouble of cutting their peat from three different levels, maintaining that this makes a difference to the malting process and thus to the quality of the final spirit. This is an obsession beyond the dictates of that brand keystone; such wilful adherence to traditional practice must drive their accountants half mad, and for no other reason I wholeheartedly commend it.

Peat cutting on Orkney looked every bit as much like hard work as it had done on Islay, so I was very pleased to observe it from the comfort of a cosy 4x4, rather than experiencing it at first hand. However, walking over the peat bogs, it was quite evident how much care was taken to eventually restore the landscape, to cut tidily and efficiently and to take some care not to disturb the birds on the adjacent reserve.

On very rare occasions, little fragments of Orkney's rich history are uncovered. One exceptional find, in the summer of 2006, was of a Bronze Age socketed axe head. Though around 3,000 years old, it was so perfectly preserved in the blanket of peat that the workers first assumed it to be an old tractor part. When first shown to professional archæologists, it was assumed to be too good to be true, but tests confirmed it to be genuine.

If you don't believe me you can read all about it, and its significance, in the journal *Vegetation, History and Archæobotany*, where the importance of the find is explained in detail over twenty pages. Top line: this was a big deal; just trust me on this.

One thing did intrigue me, though. The current theory in archæological circles is that the axe head wasn't lost, but deliberately placed as part of some kind of ritual or sacrifice to appease wetland gods or supernatural powers. There is evidence of significant early agricultural activity at Hobbister and it seems that the axe head, a valuable object of high status, was not accidentally dropped but was possibly a carefully considered votive deposition or offering.

All this peat digging goes on because Highland Park are one of the few distilleries left who maintain their own floor maltings and, again, if you pay them enough they'll let you take on as much of the tedious, back-breaking work of turning the floor of germinating barley as you can manage. I turned over a few shovelfuls just to confirm that I wasn't going to be any good at it.

However, the floor maltings here are really quite impressive and you do get a very good sense of traditional practice, exhausting as it must have been. Then there is the magnificent kiln, used to dry the grain. That's the moment at which it changes from being germinating barley, happily sprouting along in the fond belief that some kind farmer is about to plant it in his field, where it will grow strong and tall and golden and toss its hair about in the wind with all its barley friends, to being roasted over a fire and turned into malt, destined to be crushed between some rollers and have very hot water poured on it. It's a man's life, being barley.

I really like the kiln room at Highland Park. Actually, it's one of the most evocative places you can find in all of Scotch whisky. I'd rate it up there with the No. 1 Vaults at Bowmore, the Solera vats at Glenfiddich and the Springbank still room for atmosphere. It's a most agreeable place in which to stand quietly and contemplate all the loads of fuel that some poor sod has had to shovel into the firebox over the years. Plus, it's warm; often a bonus on Orkney.

But what really makes it special are two things. First, there's a priceless health and safety notice from September 1935 (10 September, to be pedantically precise). It was a Tuesday, actually, almost certainly raining, and in football news, Rochdale and Stockport County fought out a drab 1–1 draw in the English Division Three North.

The notice concerns 'Instructions for Fans etc. on Kilns' and concludes sternly, 'If it gets Stiff, Wash with Paraffin'. I've seen grown men (and women, but they should know better) dissolve into fits of suppressed laughter when they read this. It's not funny:

I've really no idea where you would get paraffin these days and a stiff regulator is not something to be trifled with.

And second, on the other side of the kiln is a solid-looking black steel door, slightly shabby, on which someone has written in chalk capitals 'SWALLOWS RETURNED 4/5/16' next to a crooked smiley face emoticon. You might expect some busybody manager or fresh-faced marketing person to come along and wipe this away, but fortunately there are no Captain Mainwaring types here and so it stays in all its glorious muddled humanity and your visit to the kiln room is all the warmer for it.

In fact, it's at this point that you know that everything about this visit is going to be great. It might be pouring with rain outside, you may be trapped in a group with some exceptionally earnest Swedes who have brought a very long list of extremely detailed questions from their whisky club, and your partner may be getting restless, which you know is going to cost you later in some as yet unspecified way, but you just know that everything is going to be great.

You know—you just know—that if they care about the swallows then they care about the peat; they care about the barley and the malting floor and the kiln; they care about the yeast and the mash tuns and the stills and the casks. Oh yes, you think, they care about the casks. And so it will prove, simply because they care about the swallows. That simple message, scrawled on a kiln door, tells you as much about the people and this place as the slick corporate video and the nicely presented shop where you also know that your credit card is shortly to sustain some serious damage, whatever your partner may think.

So, as much as anything, it's about the swallows who, quite unaware of their unlooked-for role in your reverie by the kiln, soar and swoop high above the pagodas that carry the aromatic smoke out over Kirkwall. The proper term is a Doig ventilator, so named after their inventor, the immortal Charles Cree Doig, an Elgin architect who pioneered this design to improve the efficiency of the malting process. As well as being a technical enhancement it's also a very visually pleasing structure and Doig, fortunate in both the location of his practice on Speyside and being professionally active during the great distillery construction boom of the late nineteenth century, went on to work on more than fifty distillery projects following his great innovation at Dailuaine in May 1889.

Though now largely redundant, they have become such a universally understood symbol of the entire whisky industry that today they are an almost mandatory part of a new-build design, such as at Arran, where they are completely decorative, or faithfully maintained at locations such as Ardbeg, where they no longer have any function other than to signpost the distillery.

As the distillery were patient enough to indulge my eccentric curiosity, I was taken up to the walkway at the base of the pagodas from where, if you have a head for heights, there is a fine view of Kirkwall. So from my personal inspection I can assure you that Doig's handiwork remains in good shape and is being conscientiously maintained. No one seems quite sure exactly when the ventilators were installed, but they look good for another hundred years of silent service.

I can also confirm that they ventilate very well. My tour included access to the floor on which the barley was drying and there was a steady movement of warm damp air up from the kiln, through the grain and on out to the world. It feels very strange to walk across the thick mat of barley and, though I've had this experience before in other distilleries, it's oddly affecting. Once again, quite a number of people had put in a lot of skill, time and work just for me to walk over this bed of drying grains. I hope they feel better about it now that I've drawn your attention to their efforts.

The tour, as it must, follows the standard pattern of tracing the distilling process and so I dutifully admired the mill and the mash tuns and the stills (a splendid still room, this) until being admitted to the warehouses. I always like visiting warehouses, even the enormous palletised ones that distilleries don't generally show the general public because they don't chime with the crafted image the PR folk like to portray, and not just because this is where the guide gets out some glasses and sets to work with theatrical verve. They secretly love this part because everyone is really paying attention now, with the giant bung puller and valinch to serve you some whisky. I do, for the avoidance of doubt, enjoy that bit. Except for the point where, once you've tried the whisky, the guide looks expectantly at you and asks for your profound insights. Which can, of course, only disappoint and so I always feel something of a fraud at this stage, convinced that I've talked my way here under false pretences and will shortly be found out and expelled, like some fallen angel grabbing too great a share before being ousted from the vales of heaven. But I generally get over that by the second or third cask sample.

No, what really impresses and never fails to move me is the silence and the overwhelming aroma. This is particularly affecting in traditional dunnage warehouses with their beaten-earth floors and an atmosphere accumulated over the past century or more. But even a brand-new, concrete-floored, palletised warehouse provides a rich aromatic tapestry and it's sometimes possible to walk through the ranks picking up the varied aromas of different casks, be they re-fill, ex-Sherry, former Bourbon barrels or some more exotic wine cask. Because of the intensity of the various aromas and the stillness of the air, the heady sensation can sometimes be quite overpowering.

At Highland Park the majority of the warehouses are the low-lying traditional style and it is here that the distillery prefers to store those casks that will eventually find their way into the single malt bottlings. Much is made by the distillery of their 'even-paced, cool maturation', made possible by Orkney's generally benign and temperate climate, which does not experience the variations or extremes of temperature seen, for example, on Speyside. Even allowing for marketing hyperbole it's a reasonable claim, backed up by the evidence of the weather records. 'Even-paced' and 'cool' seem an accurate enough description of Orkney as a whole, let alone Highland Park or their neighbours at Scapa.

One group of casks the guides won't be opening, however, are those containing whisky made with that Tartan barley. Tartan was once a quite popular but short-lived variety that's now more or less commercially dead. It's hard for farmers to obtain the seed, even if they wanted to, and there are many more modern varieties that are easier to grow and make them more money. But Tartan was relatively well suited to conditions in the north of Scotland, where it coped well with the short growing season and so the Agronomy Institute included it in their 2009 trials.

It's grown now on Orkney exclusively for Highland Park. The first distilling quantities were available from the 2010 harvest and early indications are that this spirit will require at least another three to four years' maturation before it can be considered for bottling. That 2010 crop, of just over fifty tonnes, is believed to be the first Orkney-grown barley malted and distilled at Highland Park since 1942, when it's thought that a small quantity of bere was pressed into wartime service, regular supplies having been severely curtailed due to the need to preserve the nation's main grain crop for food.

In order to maintain the supply chain, each of the Orkney farms will save sufficient seed for the following year's harvest, thus insulating the Orkney-grown Tartan from the wider market where the seed is, for all practical purposes, unavailable. Thus, diversity will be preserved in the event of a wider market potentially calling for early maturing malting barley, the economic viability of local farms is improved and, in due course, Highland Park will be able to release an interesting, premium product with impeccable heritage credentials and minimal food miles involved in its production. It's a casebook example of the Agronomy Institute fulfilling their goals and the academic world working hand in hand with industry (both the farming community and the distilling business) to mutual benefit. I'm looking forward to tasting it.

In the meantime, there is never any shortage of releases from Highland Park, which has been a considerable success story for Edrington, especially in tax-free markets—that's mostly airport shops as far as you and I are concerned.

Perhaps I haven't been paying attention, but I do get a little confused. There's a series named after Viking warriors—six of the bloodthirsty, axe-wielding psychopaths in all—and if you're a collector you'll want the lot. Look out for Harald, Svein, Einar, Sigurd, Ragnvald and Thorfinn and have lots of Danegeld at the ready because Thorfinn alone will cost you €1,000.

Then there was the Viking Sword release, mostly for the whisky-mad Taiwanese, and four Gods of Asgard and an Ice Edition and a Fire Edition and quite possibly a Noggin the Nog collection, though I may have imagined that one. The thing is, confusing as all this could be, it does seem to have been commercially successful. Indeed, thinking about it, a noggin of Highland Park Graculus or Grolliffe would go down very well indeed, just about now.

Though they tended to be described as 'limited editions' some of the quantities did stretch the limits of credibility. Ice, for example, a 53.9% abv seventeen-year-old, was released in an edition of 30,000 bottles at £180 and was readily accepted by the market. I may have thought the packaging contrived and clumsy, but 30,000 consumers didn't agree, so evidently the Highland Park marketing team know their audience better than I do. This, after all, is what they're paid for but, in recent months, I do get the impression that future plans are for a more restrained approach concentrating on the core range. Presumably collectors will be relieved.

It makes me happy too, as I didn't care for what I saw as a rather gimmicky and contrived set of releases. There was nothing at all wrong with the whisky itself, but the packaging and sheer quantity of the new expressions felt exploitative and the whole thing jarred. Perhaps it's just me.

With the veritable armada of cruise ships now visiting Orkney, the distillery's visitor centre is, on occasion, overrun and a victim of its own success. Apparently in the 2016 season there were a record 143 ships that berthed at Kirkwall and a lot of their passengers wanted to see the distillery.

While Kirkwall is looking very glossy and polished, not all the cruise ship passengers are necessarily well heeled. Some spend freely, while others arrive with a pretty strict budget. But little compares with the miraculous docking of the *Disney Princess*, colossus of the oceans. My sources tell me that when she puts in to harbour, the various boutique and craft shop owners get ready to book their holidays, top up their pension plans and visit the local BMW dealer. Takings have been known to jump by a third or more as a free-spending crowd descends on the town, crazed by the brutal austerity of a whole thirty-six hours without retail therapy.

There is something of a Faustian bargain that any location anywhere makes with tourism. 'Each man kills the thing he loves', wrote Oscar Wilde and he might as well have been referring to the impact of tourism on fragile cultures and places. It's one thing to visit the Ring of Brodgar, say, or for that matter the Highland Park distillery in a small and respectful group; it's altogether another in the company of a coachload of iPhone-toting tourists seeking their next Instagram picture. And if that sounds snobbish and dyspeptic then I accept the charge and I'll freely admit that I don't have any easy answers. Or really any answer at all.

I couldn't deny anyone the right to see Orkney or the pleasure to be derived from the visit, but I do regret the homogenisation of these places by the demands of mass tourism. No one fishes with grenades anymore and the rotting cars and derelict steadings have nearly all gone. A sleek, brittle sheen of prosperity has replaced the faded air that once prevailed and cottages rent for thousands per week while whisky fans queue from early morning to speculate with rare bottles.

It's not fair to suggest that Kirkwall looks like everywhere else—few places, for example, can boast a road sign warning of otters crossing between the local branches of Tesco and the Co-op, and not every Tesco has a cat with its own Facebook page—but, over the years, Kirkwall has increasingly come to resemble other places and vice versa, and its defining character and identity has been diluted. One could, sadly, say the same about almost anywhere, however, and I wouldn't be heartless enough to deny Kirkwall folk their Tesco even if more than a few rural stores have been put out of business. But those vacant premises are soon enough transformed into interesting spaces for the potters, jewellers, artists and furniture-makers who do bring such variety to the islands.

Orkney looks very well set to prosper over the next few years. Both the Highland Park and Scapa distilleries are in good hands and doing well. The nascent gin-distilling industry adds vibrancy and variety and there's every opportunity for Orkney gin to stand out as distinctive, if only by virtue of its origin, from what is becoming by the day an ever more crowded market. There must be at least 100 new gins out there now, perhaps even 101.

My wife maintains that Orkney is one of the few islands she could imagine actually living on, though I believe the standard test is to see if you can stick it through the second winter. With that behind you there's apparently some hope of a long-term residency. Perhaps one day.

If you build it, they will come

A POSTSCRIPT

THERE IS A SENSE IN WHICH THIS BOOK CAN NEVER BE completed. Not that they are building more islands but because, even as I bring it to a close, new distilleries and products are coming thick and fast, some from islands with no tradition of (legal) distilling. It's hard to keep up. No, actually, it's next to impossible, especially as some of these concerns are little more than a penny banger in a dustbin and after some initial flurry of interest on social media will quietly disappear. To keep that firework analogy going, they're up like a rocket but down like the stick. Good luck to them all though; they're great for the drinks writing community.

Someone is planning a rum distillery on Islay. I'll have to go back for that.

There are also long-standing plans for a distillery on Barra, but as the promoters have been trying to get this off the ground since at least 2005 without success, some mild scepticism is presumably justified. If a new whisky distillery hasn't been funded in the last decade of easy money, it's probably never going to happen. The last entry on the news page of their website is dated December 2014 and, with some relief, I notice that they are no longer soliciting cask sales of whisky as yet unmade. I think we'll have to file this under 'lost dreams'.

Much of the original *Whisky Galore!* Ealing Studios comedy was shot in Castlebay (Barra's principal settlement) in the summer of 1948, though the SS *Politician* was actually wrecked on nearby Eriskay. The film has just been remade. I don't hold out much hope for the new version, I'm afraid: it's almost bound to disappoint. Whoever imagined Eddie Izzard filling Basil Radford's shoes?

The Skye gin has now been launched, under the brand Misty Isle. It's craft-distilled in small batches, of course. Who would have guessed? So now there are three distilleries on Skye.

On Colonsay, the Colonsay Beverages folk have released the first of their Wild Island Botanic Gin. The Colonsay botanicals are hand-foraged, naturally. Actually, I went to Colonsay as a kid and remember seeing a school of dolphins herd fish into a small bay and take turns at swimming through them, taking mouthfuls at a time. You don't forget something like that.

Over on Tiree, the imaginatively-named Tiree Whisky Company Ltd was formed to preserve and promote the island's whisky heritage, with a view to re-establishing the connection the island once had with whisky. So naturally, they've launched a gin. The predominant flavour of Tyree Gin (for such is its rebellious designation) is achieved through locally foraged botanicals:

Eyebright, Ladies Bedstraw, Water Mint and Angelica collected from the island's rich and fertile machair ground.

But, let's face it, there wouldn't be terribly much point otherwise—though I do look forward to trying it and ticking another island off my 'to do' list.

Up in the very far north, Shetland Reel whisky and gin is being made on the old RAF airstrip at Saxa Vord on Unst, by Stuart Nickerson, once of William Grant & Sons and more recently noted for bringing Glenglassaugh back to life.

And there will be more. Meanwhile, my handwriting hasn't really improved and my wrist is sore. It's time to put my pens and my impressive collection of various inks to one side; to reach for glass and bottle (one from the islands, surely) and remember Compton Mackenzie's words:

'Ay, it's only when you haven't had a good dram for a long while that you're knowing how important it is not to go without.'[1]

Slàinte mhath!

---

1 Compton Mackenzie. *Whisky Galore*, chapter 10.

Acknowledgements

A number of people have helped bring this book together. Apart from my agent, Judy Moir, who sold the idea, my editor at Birlinn, Neville Moir (yes, they're related) has been more than patient while he waited, concealing his growing frustration (well, that's what he thinks) as deadlines slipped by.

At the various distilleries, various people helped in all sorts of ways, but particularly (in alphabetical order) I would like to thank the following for their time and kind assistance: Jason Craig (Highland Park); Alasdair Day (Raasay); Simon Erlanger (Harris); Stephen Kemp (Orkney Gin); Graham Logan (Jura); Neil MacDonald (Scapa); Euan Mitchell (Arran); Dr Nick Morgan (Diageo — Caol Ila, Lagavulin, Talisker); Sharyn Murray (Tobermory); Carl Reavey and Douglas Taylor (Bruichladdich); Mark Tayburn (Abhainn Dearg) and Jackie Thomson (Ardbeg).

Getting to and from Scotland's islands is expensive and time-consuming so I am particularly appreciative of the financial or in-kind support received from the following, again in alphabetical order, to offset the costs of travel and accommodation, without which it's doubtful this book would have been possible: Arran Distillery; Bruichladdich Distillery Co.; Burn Stewart Ltd; Chivas Brothers Ltd; Diageo plc; Highland Distillers Ltd; and Isle of Harris Distillers Ltd.

All of the distilleries concerned were more than kind in providing product samples where requested, for which I thank them. I apologise for the paucity of sycophantic tasting notes and promise to try harder another time (or, in other words, just keep the samples coming).

A note on the photographs

All images are ©Ian Buxton, 2017 except as indicated. Most of my photographs were taken as personal aide memoires, a reminder of a specific time or place, with no intention that they would appear in the book. However, as it became clear to me that I did not want to use the artfully photoshopped, carefully curated and PR-industry-mediated official shots that the various distilleries' marketing teams are ever anxious to provide, the very artlessness of my 'snaps' came into sharper focus and, for all their technical and compositional shortcomings, seemed to offer a more honest complement to the text and a less sanitised view of the islands. I saw something of the same quality in the third-party images.

The author and publisher acknowledge with thanks the following for the use of photography. All copyrights are acknowledged. *Sam's Grave* North Harris Trust. *Tobermory Distillery* Distell International Ltd. *Willie Cochrane* Whyte & Mackay Ltd. *Yellow Submarine* Carl Reavey. *Bunnahabhain Cottages* Alistair McDonald. *Bowmore Peat Cutters* Bowmore Distillery.